USING TOOLPACK SOFTWARE TOOLS

ON INFORMATION SCIENCE

A series devoted to the publication of courses and educational seminars given at the
Joint Research Centre, Ispra Establishment, as part of its education and training program.
Published for the Commission of the European Communities,
Directorate-General Telecommunications, Information Industries and Innovation,
Scientific and Technical Communications Service.

The publisher will accept continuation orders for this series which may be cancelled at any time and
which provide for automatic billing and shipping of each title in the series upon publication.
Please write for details.

USING TOOLPACK SOFTWARE TOOLS

*Proceedings of the Ispra-Course held at
the Joint Research Centre, Ispra, Italy, 17-21 November 1986*

Edited by

A. A. POLLICINI

Joint Research Centre, Ispra, Italy

KLUWER ACADEMIC PUBLISHERS

DORDRECHT / BOSTON / LONDON

Library of Congress Cataloging in Publication Data

Using Toolpack software tools / edited by A.A. Pollicini.
      p.   cm. -- (Ispra courses on information science)
   Bibliography: p.
   Includes index.
   ISBN-13:978-94-010-6883-3   e-ISBN-13:978-94-009-0879-6
   DOI: 10.1007/978-94-009-0879-6

   1. Toolpack (Computer program) 2. Mathematics--Data processing.
 3. Computer software--Development.   I. Pollicini, A. A. (Aurelio
 Alberto), 1938-   . II. Series.
 QA76.95.U75 1988   88-31613
 510'.28'551--dc19   CIP
 ISBN-13:978-94-010-6883-3

Commission of the European Communities Joint Research Centre Ispra (Varese), Italy

Publication arrangements by
Commission of the European Communities
Directorate-General Telecommunications, Information Industries and Innovation, Scientific and
Technical Communications Service, Luxembourg

EUR 11924
© 1989 ECSC, EEC, EAEC, Brussels and Luxembourg
Softcover reprint of the hardcover 1st edition 1989

LEGAL NOTICE
Neither the Commission of the European Communities nor any person acting on behalf of the
Commission is responsible for the use which might be made of the following information.

Published by Kluwer Academic Publishers,
P.O. Box 17, 3300 AA Dordrecht, The Netherlands.

Kluwer Academic Publishers incorporates the publishing programmes of
D. Reidel, Martinus Nijhoff, Dr W. Junk and MTP Press.

Sold and distributed in the U.S.A. and Canada
by Kluwer Academic Publishers,
101 Philip Drive, Norwell, MA 02061, U.S.A.

In all other countries, sold and distributed
by Kluwer Academic Publishers Group,
P.O. Box 322, 3300 AH Dordrecht, The Netherlands.

to my wife Bruna

Contents

Tool Writing – W. R. Cowell   **251**

Open Forum   **277**

The Future of Toolpack / 1 – I. C. Hounam   **297**

Trademark Acknowledgements

Ada is a registered trademark of the U.S. Government – Ada Joint Program Office

Apollo is a trademark of Apollo Computer Incorporated

CYBER, NOS, NOS/VE are trademarks of Control Data Corporation

DEC, VAX, VMS are trademarks of Digital Equipment Corporation

IBM is a registered trademark and CMS, MVS, TSO, VM
    are trademarks of International Business Machines Corporation

NAG is a trademark of The Numerical Algorithms Group Limited

TEX is a trademark of the American Mathematical Society

UNIX is a trademark of AT&T Bell Laboratories

Foreword

I am very pleased to write these few brief paragraphs introducing this book, and would like to take this opportunity to attempt to set the Toolpack project in an appropriate historical context.

The Toolpack project must be considered to have actually began in the Fall of 1978, when Prof. Webb C. Miller, at a meeting at Jet Propulsion Laboratories in Pasadena, California, suggested that there be a large-scale project, called Toolpack, aimed at pulling together a comprehensive collection of mathematical software development tools. It was suggested that the project follow the pattern of other "Pack" projects, such as Eispack, Linpack, and Funpack which had assembled and systematized comprehensive collections of mathematical software in such areas as eigenvalue computation, linear equation solution and special function approximation. From the beginning it was recognized that the Toolpack project would differ significantly from these earlier "Pack" projects in that it was attempting to assemble and systematize software in an area which was not well established and understood. Thus it was not clear how to organize and integrate the tools we were to collect into Toolpack. As a consequence Toolpack became simultaneously a research project and a development project. The research was aimed at determining effective strategies for large-scale integration of large-scale software tools, and the development project was aimed at implementing these strategies and using them to put high quality tools at the disposal of working mathematical software writers.

We at the University of Colorado, Boulder felt privileged to take the technical lead during the formulation of the early Toolpack architectural notions and later system design. Realizing that the discipline of software engineering was a young one and that the enterprise of software tool development was poorly organized and understood, we took as our central premise the need to integrate our tools in a highly flexible and extensible way. We believed that the only constant in this area would be the continuing need to upgrade tools and effectively assimilate new ones. Recognition of this need led rather directly to the key notions of modular tools and well-defined software objects – intermediate and final. Our early prototypes supported our beliefs that these architectural notions were robust and could be implemented effectively, thereby supporting a well-integrated, yet efficiently extensible, collection of tools.

We exported these ideas and prototypes to Nag Ltd., Oxford England, in 1983-1984, entrusting to their capable hands the development and distribution of an industrial-strength version of Toolpack. The success of their efforts is attested to by the community which has formed and which is now requiring this book.

At Boulder we are also gratified by this growing acceptance of Toolpack because we believe that it reflects well upon the architecture and design notions which went into the original Toolpack prototypes. As users continue to have success in effectively assimilating new tools and altering older tools, we increasingly obtain validation of our original notions and indications of how they might be improved. The preliminary indications are that many of our early ideas have indeed been validated, but that they also indicate important directions for significant improvements.

Nine years after the inception of the Toolpack project we find that the field of software tool development and integration (now called software environments) is still not well enough organized and understood. We do find, however, that a number of other environments and tool systems have built effectively upon some of Toolpack's architectural notions, suggesting that this project has indeed advanced progress in this field. Moreover, our experience with Toolpack has confirmed in our minds that tools are best thought of as very high level operators, operating on very high level software objects. This has led to the notion of software process programming which underlies, for example, Arcadia – a research and development project whose aims are to explore software tool integration notions and to move the successful ideas to actual practice.

Thus Toolpack has served as a model of how research and development can be advanced effectively in parallel and how each advance that is used in practice paves the way for further research. The Toolpack community can be justifiably proud of its rôle in both advancing the state of the practice in exploiting the power of software tools and the state of the art in knowing how best to effect this. On the landmark occasion of the publication of this book I extend my most sincere thanks to all who helped in the early stages of this work. I am especially indebted to Geoff Clemm and Stu Feldman. I also extend my heartiest congratulations to those who have so successfully carried the work forward, most notably Steve Hague, Brian Ford, Allan Shafton, Wayne Cowell, Bob Iles, and Malcolm Cohen.

Leon Osterweil
Boulder, Colorado
24 August 1987

Preface

Behind any achievement there is always a story.
This is also the case for the publication of this book on "Using Toolpack Software Tools".

It may be interesting to tell the reader some fragments of that story before entering the technical matter of the book itself.

It is, of course, a story of facts and people which began in 1981, when at JRC Ispra we became aware of the Toolpack project. It may be worth quoting a couple of sentences from two letters of that period:

> *" Dear Dr. Cowell,*
>
> *during the course ot this year we have read and heard various interesting things about the Toolpack project.*
>
> *...*
> *... we certainly would have more than a passing interest in providing manpower and facilities in the testing and refinement phase of the project."*

wrote Martyn Dowell of JRC Ispra, on 19th October 1981.

> *" Dear Dr. Dowell,*
>
> *...*
> *... We ... expect to select a few sites for field testing sometimes next year, ...*
> *I suggest that we discuss further a possible rôle for your organization after ... we have moved closer to a field test version."*

answered Wayne Cowell from Argonne, on November 6th.

These documents – *alpha* and *omega* of the admission of JRC Ispra to an active rôle in the promotion of the Toolpack environment – established the first contact between two people I have to acknowledge publicly. Thanks to Martyn, with whom I have been working in a friendly and professional symbiosis since 1979, and to Wayne who has had, since the origin, leading responsibilities within the project, I had the opportunity of entering the fascinating world of Toolpack.

The idea of enlarging that world to the wider Fortran community went through education and was actually embodied into the organization of the first international workshop on using Toolpack tools, held at Ispra in the week 17th to 21st November 1986. The person who worked most for drafting the programme of the workshop and for co-ordinating integration and availability of the software was Bob Iles whose efforts I gratefully acknowledge. To the benefit of all workshop participants, from Nag Central Office in Oxford, he played the rôle of a discreet and unreplaceable prompter to all the actors on the scenes of the workshop.

Not only over men, but also over institutions relied the effort of gathering at Ispra the experience of designers, developers, implementors and distributors of Toolpack tools. To mention only two, I wish to thank The Numerical Algorithms Group, Oxford, in the person of its deputy director Steve Hague who contributed both personal and corporate support to the initiative, and the Ispra Educational service EFIS, in the person of its head Rolf Misenta who hosted the workshop in the 1986 programme of Ispra-Courses, a 12 year established tradition in our establishment.

Now the materials presented, developed and discussed during the workshop are taking the form of a book which is typeset using the TEX system. This final step is another experimental achievement on our site, which is progressing with the design and typographical layout of Filippo Franchini and with the patient editing of Paolo Gaglione. Their collaboration is worthy of mention.

Aurelio Alberto Pollicini

Ispra, Italy

15th November 1987

Introduction

Aurelio A. Pollicini
Joint Research Centre
Ispra, Italy

".... it takes all the running you can do, to keep in the
same place. If you want to get somewhere else, you
must run at least twice as fast as that!"

— Lewis Carroll,
Through the Looking-Glass, ch. II, 1872

All is running faster and faster in computer science. Indeed, from the advent of
high level programming languages to the present time, computer programming had
changed from a free-style creative activity to an engineered production of problem-
solving wares: *the software*. Software engineering is faced with activities such as
the development, the repair and the maintenance of software as well as with the
management and the control of these activities. The trend is to improve both
productivity rate and product reliability by enhancing methodologies and by applying
automatic tools to all the phases of the entire software life-cycle. Software tools,
rather than isolated and ad-hoc utility programs are ever more commonly considered
as inter-related components of a co-operating environment: *the programming support
environment*. In the course of the project sponsored by the US Department of Defence
for the development of the programming language Ada, it was pointed out that other
important functionalities must supplement the provision of language processing, thus
developing the requirements for an Ada Programming Support Environment (APSE).
Moreover, interesting analogies were found between APSE components and the phases
of the software life-cycle. On one hand, this perception promoted studies for an Ada-
based System Development Methodology, on the other hand, it helped to promote the
idea that programming environments are a common requirement for the use of any
programming language.

If we think of an environment around the Fortran language, we have to split the matter
into two fields: programming support environments and the Fortran world.

Because of its wide scope, a programming support environment may be seen as a
number of layers communicating through an information base. Each inner layer being

1

A. A. Pollicini (ed.), Using Toolpack Software Tools, 1–5.
© *1989 ECSC, EEC, EAEC, Brussels and Luxembourg.*

the basis for the surrounding one. (So we have KAPSE and MAPSE defined within APSE.) Layers at different levels provide different coverage of tool support. At the lowest level we have tools orientated to program execution whilst at the next higher level we have tools orientated to program manipulation. At even higher levels there are tools orientated to life-cycle support and project support. If the toolset lacks these higher levels we consider it to be providing a *program development system* rather than a programming support environment.

Regarding the Fortran world, which is only a small fraction of the programming language area, we must consider the intrinsic aspect of Fortran survival. Beyond any statement of "practical usefulness" for explaining its past success, there is an ever growing identification of Fortran with the exploitation of vector and parallel computers. Therefore, it is quite likely that this pragmatical rather than systematic language will be in effective use for a long time to come. In the wider context of computer programming, procedural languages trend towards extending the levels of abstraction for the enhancement of the design, development and maintenance of reliable software. How Fortran evolves is mainly traced out by the process of revision of the current standard Fortran 77. The proposals for the draft standard – known as Fortran 8x – aim at a language both upward compatible with Fortran 77 and open to advanced concepts of modern language design.

When one is talking about trends one talks of opinions that may be not generally shared. Nevertheless, it seems justified to say that there are common concepts behind programming languages and that the functionalities which implement such concepts are migrating from language to language so that different languages tend to converge to a global set of general purpose features, though varying considerably in syntax. In other words, rival languages such as Algol and Fortran have spawned descendants with an ever increasing common base of functionalities. For instance, Pascal and Fortran 77 are less different than their ancestors and, functionally, Ada and Fortran 8x seem still closer.

As functionalities spread out, so the effort of generalizing programming support environments to include Fortran is neither impossible because of deficiencies of the language, nor useless because of lack of perspectives.

It is within this contextual generalization that we have to consider the intents of the Toolpack project, which aimed at – professor Osterweil said in 1983 [45] :

" ... *a toolset ... specifically designed to be highly flexible and extensible* "

" ... *a portable software system ... operated as a system for the manipulation and management of the large, complex and multifaceted object which is the software program under development* "

" ... *to offer the Toolpack user a particularly powerful yet natural interface to the toolset* "

" ... *by forming and manipulating various mental models which can be thought of as abstractions or projections of the source code representation* "

By its design goals, and in consideration of the wide scope of the toolset, we can say that Toolpack ranges at the top of program development systems, with some of the attributes of a programming support environment. Additionally, the requirement of being itself transportable through diverse computer systems makes Toolpack a catalyst of cohesion within the Fortran 77 software industry.

Toolpack software was publicly released in Spring 1985. After 20 months of experience over a number of sites, almost in concomitance with the distribution of the second release of Toolpack/1, a workshop was organized at JRC Ispra, attended by active project members, installers and/or users of the first release and prospective Toolpack users. (Names and addresses are given below.) The worst mistake that could have been embedded in the message conveyed by this workshop would have been to stress the notion of tool fragmentation without penetrating with equal emphasis the notion of fragment integration. Indeed, the functionalities implemented by individual fragments, from the usage point of view, must be considered as conjugated as the teeth of two cogwheels. We hope such a mistake was avoided and also that the material presented in the 18 chapters of this book gives the reader the flavour of a fully integrated development environment.

The first part of the book is devoted to general overviews of the Toolpack project, the structure of Toolpack/1, the portability base TIE, the tool integration framework and the contents of the tool suite.

The second part describes a number of individual tool fragments, as well as practical ways of integrating them, for achieving a variety of common tasks such as: editing source with intelligence of Fortran syntax, standardizing the declaration part of a program, source tidying, standard conformity checking, rearranging unstructured control-flows, changing precision, comparing files of data, instrumenting a program and issuing reports.

The third part presents a wider insight into Toolpack. It gives an overview of the multidimensional space of Toolpack/1 installation, of tool suite extension by developing TIE conforming tools, of confrontation between today's experiences and tomorrow's expectations, and of the short term evolution of Toolpack itself.

The wish that any reader can become a "friend of Toolpack" has guided the work of all who contributed to the publication of this book.

List of Lecturers

Malcolm J. Cohen   The Numerical Algorithms Group Ltd
Wilkinson House, Jordan Hill Road
Oxford OX2 8DR, United Kindom

Wayne R. Cowell   Argonne National Laboratory
9700 South Cass Avenue
Argonne, Illinois 60439, USA

Martyn D. Dowell   Joint Research Centre
Ispra Establishment
I-21020 Ispra, Italy

Stephen J. Hague   The Numerical Algorithms Group Ltd
Wilkinson House, Jordan Hill Road
Oxford OX2 8DR, United Kingdom

Ian C. Hounam   The Numerical Algorithms Group Ltd
Wilkinson House, Jordan Hill Road
Oxford OX2 8DR, United Kingdom

Aurelio A. Pollicini Joint Research Centre
Ispra Establishment
I-21020 Ispra, Italy

List of Workshop Participants

Eugène   Audin   Électricité de France
1, Avenue du Général De Gaulle
F-92140 Clamart, France

Pannalal Bose   Nordostschweizerische Kraftwerk AG
Parkstrasse 23, Postfach
CH-5401 Baden, Switzerland

Stephen Brazier   Prime Computer Λ
Walkern Road
Stevenage, Herts SG13QP, United Kingdom

Martin A. Broom   University of Kent
Computer Laboratory
Canterbury, Kent, United Kingdom

Philippe Defert   E.S.O./ST-ECF
2, Karlschwarzschild Strasse
D-8046 Garching, Federal Republic of Germany

Ulrich Detert   KFA Jülich Gmbh
Postfach 1913
D-5170 Jülich, Federal Republic of Germany

Raoul M. Doron   Joint Research Centre
Ispra Establishment
I-21020 Ispra, Italy

E. De Boer   N. V. Kema
Utrechtsweg 310
NL-6812 AR Arnhem, The Netherlands

Hervé P. De Sadeleer   Joint Research Centre
Ispra Establishment
I-21020 Ispra, Italy

Thierry Engels   Solvay & Cie
310 rue de Ransbeek
B-1120 Bruxelles, Belgium

Jeffry C. Kvam   The Numerical Algorithms Group Inc
1101 31st Street, Suite 100
Downers Grove, Illinois 60515, USA

Minaz Punjani   University of London Computing Centre
20 Guilford Street
London WC1N 1DZ, United Kingdom

Ulrich Spieker   Joint Research Centre
Ispra Establishment
I-21020 Ispra, Italy

Silvio Valente   Politecnico di Torino
Corso Duca degli Abruzzi 24
I-10129 Torino, Italy

Aristide V. Varfis   Joint Research Centre
Ispra Establishment
I-21020 Ispra, Italy

Gilbert Verstappen   SCK/CEN Dept. Ceramic Research
Boeretang 200
B-2400 Mol, Belgium

Yuthi Yem   Soc. de Réalisation
en Systémique Industrielle
Chemin du Pré Carré
F-38240 Meylan, France

Part I
The Toolpack Environment

"Je suis jeune, il est vrai; mais aux âmes bien nées.
La valeur n'attend point le nombre des années."

— Pierre Corneille,
Le Cid, act II, sc. 2, 1636

The Toolpack Project

Stephen J. Hague
NAG Ltd
Oxford, U.K.

"It has, of course, been impossible for me to keep "two years ahead of the state of the art," and the frontiers mentioned above will certainly change. I have mixed emotions in this respect, since I certainly hope this set of books will stimulate further research, yet not so much that the books themselves become obsolete!"

— Donald E. Knuth,
Preface to "The Art of Computer Programming",
October 1967

Abstract. In this paper we describe the background to the Toolpack project from its inception in 1979 to the launching of public distribution services by NAG in 1985 and 1986. We also discuss basic concepts in using software tools and give examples of how these tools can be used to automate software processing activities.

1. Introduction

Our attention in this publication is focussed on the use of software tools in the second release of the Toolpack/1 Fortran 77 software tools suite by NAG in November 1986. This followed the first release of Toolpack/1 which was made available in February 1985. That first release was the result of an international collaborative activity (the Toolpack project) started in 1979. We begin by describing the background to that project.

1.1. Background to the Toolpack Project

The Toolpack project, which formally ended in 1984, was supported in the U.S.A. by the National Science Foundation and the Department of Energy, and in the U.K. by

9

A. A. Pollicini (ed.), Using Toolpack Software Tools, 9–22.
© 1989 ECSC, EEC, EAEC, Brussels and Luxembourg.

the Science and Engineering Research Council. The conceptual design of Toolpack as an integrated suite of tools, and the construction of the prototype system based on that design, was carried out by Professor L. J. Osterweil and colleagues at the University of Colorado. Professor Osterweil and Dr. S. J. Hague served as technical co-chairmen of the project, which was co-ordinated by Dr. W. R. Cowell of Argonne National Laboratory. Toolpack/1 was assembled by R. M. J. Iles, I. Hounam and M. J. Cohen of the NAG Central Office and prepared for release in conjunction with Argonne National Laboratory. Other organisations involved in the original Toolpack project were Bell Communications Research, Jet Propulsion Laboratory, University of Arizona and Purdue University.

1.2.   Original Aims

The original project aims were twofold:

(1) To provide a suite of tools to assist the production, testing, maintenance and transportation of medium-sized mathematical software projects written in standard-conforming Fortran 77.

(2) To investigate the development of extensible programming support environments built around integrated tool suites.

In this context, "medium-sized" software projects were considered to be those of about 5000 lines in length, being created by up to three programmers. A need to support existing software projects of up to twice that size was also recognised.

Several other important factors were taken into account:

(1) The tools should operate on standard-conforming Fortran 77 programs.

(2) The facilities should be biased towards interactive use, but with a batch mode capability.

(3) Facilities should be provided to manage and develop associated documentation, test data and results.

(4) The tool suite should be portable across a wide spectrum of host systems, so as to allow as many users as possible to gain experience of its facilities.

Above all, it was recognised that the tool capabilities must be powerful, reasonably efficient and easy to use. Given the diverse nature of the Fortran community, it was recognised that the relative importance of these characteristics would vary from one site to another. The desire to meet these varying requirements further emphasized the need for flexibility. The Toolpack project did not set out to devise or support any specific programming methodology, but to provide a versatile and extensible set of facilities which could be configured either to perform individual tasks or to support such methodologies.

The choice of Fortran (specifically Fortran 77) as the subject language was a reflection of the primary interests of the participants in the Toolpack project. Since the early

1970s, mathematical software developers such as NAG appreciated the value of software tools and had experience of using various, independently produced Fortran (66) programming aids. These developers and computer scientists interested in supporting the scientific computing community launched the Toolpack project in 1979. Given the continuing dominant position of Fortran in that community (and the advent of Fortran 77 compilers around that time), Fortran 77 was an un-disputed choice. The question of which extensions, if any, of Fortran 77 to support was the subject of some debate. Eventually a modest extension of the standard language definition was adopted – see Section 4 of the Chapter "The Toolpack/1 Tool Suite".

1.3.   Promoting Awareness and Identifying Requirements

An important objective of the Toolpack project was to educate Fortran programmers to expect better support for software development. Until quite recently the only tools likely to be available to the Fortran programmer have been an editor, compiler and possibly a run-time debugging system (though this might well not operate at the source text level). More powerful tools for Fortran had been developed on a largely ad hoc basis, each according to the specification of its own designer. For computer users unable to develop their own tools, their efforts to establish an effective set of programming support tools were likely to encounter various difficulties. Even if suitable tools could be obtained, they were likely to require separate installation, they might not support the same dialect of Fortran, and would probably exhibit other differences in areas such as the user interface and the quality of documentation, for instance. In particular, users attempting to mount tools on more than one host system were unlikely to find that similar facilities were available for each of the systems.

The Toolpack project was an ambitious attempt to make substantial progress towards alleviating, if not eliminating, the difficulties described above. Though the original Toolpack project has ended, it is premature to judge its effect. The dissemination and promotion of software tools for Fortran programmers is an activity that must be assessed over a period of several years, particularly given the limited resources available to co-ordinate such efforts, and the rather conservative nature of the Fortran user community. It is also crucial to recognise the need for the tools to evolve; in the light of user response to the first release of Toolpack/1 software, a significant number of improvements and enhancements have been made in the second release. There is every prospect that this latest release will reach a wider audience within the Fortran community and thus further stimulate user awareness. Even with our limited experience so far, it is evident that, once users have access to reasonably efficient and reliable software tools, they appreciate the value of the tools and may well identify potential rôles for further tools. Since Toolpack/1 contains tool writing facilities, it will often be quite feasible to construct these additional tools. Users will rightly come to expect that the same level of programming support facilities should be available on all the systems upon which they develop or maintain software. Thus, Toolpack has a distinct two-way educational rôle in promoting awareness amongst the user community and, in return, learning more of the support services that users would actually wish to have. Users should be encouraged to expect that facilities equivalent or indeed superior to those in Toolpack/1 should be available, as a matter of course, as part of their host Fortran compilation system.

2.   Toolpack/1 Services

2.1.   Relationship of Release 2 to Release 1

The first release of Toolpack/1, which has been distributed to over 400 sites, provided Fortran 77 programmers with a suite of consistent tools that could be used to manipulate programs conforming to the ANSI Fortran 77 standard [1], their test data, results and documentation. The use of these tools to mechanise elements of the software life cycle can lead to a greater consistency and reliability in the resulting code. In addition to these tools, Toolpack/1 led to the provision of several other facilities:

(1) a portability base, TIE (Tool Interface to the Environment) designed to allow tool developers to design and produce portable tools within the Toolpack/1 framework,

(2) access functions to allow the tool developer to read, manipulate and write intermediate objects for exchange with other Toolpack/1 tools,

(3) supplementary libraries of routines providing enhanced facilities for string, table, list, pattern and screen handling,

(4) an experimental object-orientated programming support environment for assessment and possible further development.

These additional facilities mean that Toolpack/1 also supports the programmer wishing to develop portable tools (including those for the manipulation of Fortran 77 programs) or experiment with programming environments.

Release 2 supersedes Release 1 of Toolpack/1. The definition of TIE (the portability base) has remained unchanged but the tools and supplementary libraries have been modified, enhanced and expanded. Existing implementations of TIE can support the Release 2 tools and supplementary libraries. The enhancements include:

(1) all bug fixes and performance enhancements described in Release 1 update notices,

(2) new tools, in particular a static semantic analyser and a Fortran portability verifier,

(3) the ability of tools to process an enhanced dialect of Fortran 77,

(4) improved and updated documentation.

The current contents of the available Release 2 services are summarised in **Appendix A**.

2.2.   Public Access/Domain

The first release of Toolpack/1 was the result of a collaborative research project to study the design and production of programming support tools and so to stimulate interest in and the development of properly supported programming environments. Because of the nature of the project funding, the resulting software was in the public domain. NAG played a major rôle in developing and assembling Release 1, and agreed to provide a public distribution service, the aim of which was to enable Fortran 77 programmers to make sample use of the tools developed during and after the project. The Release 1 service terminated on 1 September 1986 and no further Release 1 tapes are available from NAG.

Continuing in the spirit of the public access/domain philosophy, NAG co-ordinated the assembly of Release 2 and has launched extended distribution and information services for this latest release. Considerable care has been taken in preparing Release 2 for general use; indeed many Release 2 tools are in regular use within NAG in the development of software products. However, Toolpack/1 (Release 2) remains in the public domain; it is not a fully supported commercial product. In due course, enhanced programming support facilities may be available from NAG as part of a supported commercial service but Toolpack/1 (Release 2) is not a NAG product.

Possible future developments are discussed in the Chapter "The Future of Toolpack/1".

3.   An Introduction to Software Tools

This section is devoted to an explanation of the nature and rôle of software tools. It is intended primarily to be read by the less experienced Fortran programmer who is unfamiliar with terms like *software tools* and *programming aids*. Other readers familiar with these concepts may prefer to proceed to Section 4 of this chapter for a description of the use of Toolpack/1 tools.

3.1.   Examples of Software Tools

Toolpack is a suite of software tools designed to support the Fortran programmer. In this context, a "software tool" is a utility program that may be used to assist in the various phases of constructing, analysing, testing, adapting or maintaining a body of Fortran software. Alternative terms used to refer to such programs are programming aids and *mechanical aids*. We will henceforth usually refer to these programs simply as "tools". Typically, the input to such a tool is the user's Fortran software. That software may be part of a program, a complete program or indeed a suite of programs. The tool reads the software as input data, processes it and produces output that may have one or both of the following forms:

(1) A report that gives an analysis of the input program, e.g. a summary of the types of statements used. A tool that produces this form of report without requiring the user program to be executed is called a *static analyser*.

(2) A modified version of the input program. In this case, the processing tool is a *transformer*. An example of such a tool is a formatter which improves the appearance of the user's program. The Toolpack/1 formatter is called "polish".

In some cases the input may not be a program as such, but could be an associated or derived body of information, e.g. test data, documentation, or a report based on information generated by a previously applied tool. Tools that assist directly in the preparation of documentation are usually called *documentation generation aids*. These and other tools serving utility functions all have an important rôle to play and so, even if they do not process the user program directly, they are still regarded as programming aids.

Further examples of software tools of interest to the Fortran programmer include:

(1) a text editor, with Fortran 77 oriented features, which is useful for creating a new Fortran program and then modifying it,

(2) a transforming tool that standardises the declarative part of a Fortran program,

(3) an instrumentor that modifies the program by inserting monitoring and other control statements. The instrumented program is then compiled and executed and data is gathered that is used to generate reports. Execution of an instrumented program is an example of *dynamic analysis*.

3.2.   Mechanisation Through the Use of Tools

Examples of each of the tools mentioned in the previous section can be found in the contents of the second release of Toolpack. The list of facilities in that release, summarised in Appendix A, is by no means exhaustive; there are many more tools which could be of considerable use. These could be written to conform to Toolpack/1 construction guidelines using the facilities outlined in the Chapter "Tool Writing".

Readers unfamiliar with the use of software tools may not be convinced that these efforts to provide enhanced support for Fortran programmers are either necessary or worthwhile. After all, they might argue that the Fortran language has survived for over 30 years without the availability of such tools in widespread use. They might also wonder why terms such as "software tool" and "programming aid" have become fashionable only recently, and are still not part of some Fortran teaching courses. More fundamentally, they might ask why the major vendors of Fortran compilation systems have apparently paid little heed to the desirability of providing various program analysers and transformers.

To attempt to give full answers to the above questions would be inappropriate in this introductory chapter since numerous technical and other issues are involved. However, the upsurge in interest in Fortran software tools can be ascribed mainly to the experiences of numerical software producers during the 1970's. In their efforts to build soundly constructed, well tested, portable products, these producers came to appreciate the potential value of using software tools and the associated benefits of mechanising parts of their particular software production cycles.

The main potential benefits claimed for mechanisation are:

(1) *Reliability*: if a thoroughly tested tool operates on a valid user program, then the outcome should also be predictably trustworthy.

(2) *Economy*: if a tool is to be applied repeatedly, then the costs of its development and usage may be significantly less than those of performing the task by hand, given a sufficient scale of operation.

(3) *Consistency*: related to the notion of reliability, there is the prospect that programs processed by a tool will be more uniform in appearance and construction than if they were changed by hand. This makes programs easier to understand and to maintain.

(4) *Flexibility*: if a program is to be transformed by a tool whose effects can be readily altered, then such a tool gives scope for producing alternative forms (which may be useful both for experimental purposes or for the management of variant forms of the program to be used in different contexts).

There are therefore several strong arguments for considering the use of software tools. Having outlined these advantages, however, it is necessary to mention possible drawbacks, particularly those that the less experienced Fortran programmer might encounter. Perhaps the two most important cautionary notes to make are:

(1) *Over-reliance on the outcome*: the user may assume that it is unnecessary to make any kind of visual or computational check on a transformed program before, for instance, sending copies of that program to colleagues elsewhere.

(2) *Underestimation of required resources*: the exact computational costs of using software tools will vary considerably (and in the case of Toolpack/1 will depend critically on the efficiency of the implementation of the underlying portability software). Users should not underestimate the storage and processor requirements of using software tools.

However, despite these reservations Toolpack/1 will be of practical benefit to many users and it will give a foretaste of the kind of comprehensive, integrated tool suite that ought to be at the programmer's disposal.

3.3.   Basic Assumptions

To use Toolpack tools, it is not necessary to have an intimate understanding of their internal operations, although for the interested reader some details are given in [17]. Both in that publication and in the specifications of the individual tools, it is assumed that the reader is familiar with certain basic assumptions about the way that the tools operate and the several forms of program representation. In this section we introduce those assumptions in general terms.

Taking first a general case, assume that a software tool, T, operates on a user program in source form, which we will denote by P. Let us assume that the tool produces an output program in source form, to be denoted by P'. In order to carry out its operations efficiently and reliably, the tool T must first "recognise" the program P. This recognition involves at least a *lexical analysis* of the program, which means that the individual language *tokens* (sometimes called *atoms* or *lexemes*) are delimited and identified.

A tool which carries out a lexical analysis is called a lexer, or sometimes *scanner* since the analysis involved is a preliminary scan of the program.

Lexical analysis guarantees only that the individual elements of the program are valid; it does not ensure that the code conforms generally to the grammatical (syntactical) rules of the appropriate language definition (Fortran 77 in the case of Toolpack). In order to ensure that the program is grammatically correct, a further checking operation after lexical analysis is carried out. This operation is known as *syntax analysis* or *parsing*. The syntax analyser (or "parser") takes the token stream as input, checks that it is grammatically valid according to the appropriate language rules, and produces a *parse tree* as output. The parse tree reflects the correct syntactical structure of the program, and each statement of the program corresponds to a distinct subtree in the parse tree.

Whilst performing its analysis, the parser encounters program elements with names or values, e.g. the identifier named X, or the constant 1. It stores this information in a "symbol table", for ease of reference in later processing. In fact, the symbol table may exist after lexical analysis (depending on the particular lexical tool used), so in that case the parser adds more information to an existing table (e.g. it will distinguish between the arithmetic constant with value 10, and the label 10, whereas the lexer may not).

Syntax analysis checks that the program conforms to the formal grammatical rules of the language but those rules may not cover all aspects of the use of the language, particularly with regard to consistency within the program and, most crucially, the significance of each instruction, i.e. the "semantics" or meaning of the program.

The task of checking for such inconsistencies and contraventions is called *semantic analysis*. In the process of performing these checks, more information is gathered about various aspects of the program, and this additional data is usually stored in an *attribute table* associated with the symbol table produced by the syntax analyser.

The output from a full semantic analysis may be voluminous, and much of the information may not be directly relevant to the input requirements of a subsequent processing tool. This implies that storage and processing resources could be unnecessarily expended whilst producing unwanted information. Therefore, the strategy adopted in Toolpack/1 is to incorporate into the parser a limited degree of semantic analysis, producing a modest amount of information likely to be of use to most other tools. If any of those tools has additional analysis requirements, these are fulfilled either by the tool itself or by a separate semantic analyser.

These attributes can then be added to by the semantic analyser. An additional feature of the Toolpack/1 semantic analyser is that it will check conformance to the Fortran 77 standard, providing warnings of the use of extensions and invalid constructs.

3.4.   Summary of the Recognition Process

To summarise the discussion so far, we observe that the process of "recognising" the program, **P**, involves lexical, syntactical and semantic analysis and produces two distinct "sub-source" forms of program representation (the semantic analyser merely enhances the attribute information derived by the parser) as illustrated in Figure 1.

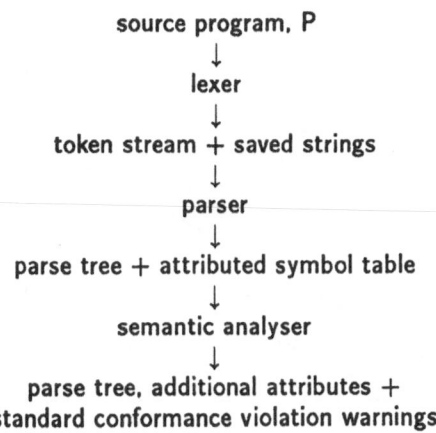

source program, P
↓
lexer
↓
token stream + saved strings
↓
parser
↓
parse tree + attributed symbol table
↓
semantic analyser
↓
parse tree, additional attributes +
standard conformance violation warnings

Figure 1. Program Analysis and Forms of Program Representations

3.5.   The Overall Transformation

To complete our description of a source-to-source transforming tool, **T**, we must now consider how the eventual source output form is produced. If **T** operates on the token stream, then it must incorporate a utility which converts the modified token stream back to source form. This utility is in fact the "polish" tool mentioned earlier. It converts Fortran keywords from an internal coding form for tokens to character strings and inserts spaces and newline characters where appropriate. This operation is sometimes called *unscanning* since it is the opposite of the initial lexical analysis phase, known as scanning.

If the tool **T** operates on the tree form, then before unscanning, the revised tree must be "flattened" into a token stream. This essentially involves traversing the leaves of the parse tree but also must take into account any modifications made to the symbol/attribute tables. The flattening may take place as part of the tree transforming operation within **T**, or as a separate phase.

The operation of our hypothetical tool, **T**, therefore could involve one of several transition paths between **P** and **P'**, as illustrated in Figure 2.

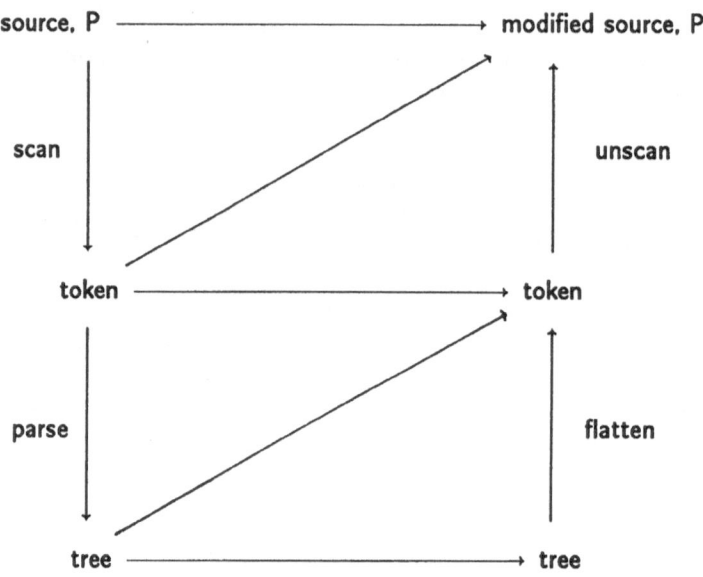

Figure 2. Example of Transition Paths

Further details of the lexer, parser, semantic analyser, polisher and other related tools will be found in the chapters of Part II.

4.   Examples of the Use of Toolpack/1 Tools

In this section two examples of the uses of Toolpack/1 Release 2 tools are given. The first example describes how the tools may be used during the development of a software product, the second example describes the uses of the tools in NAG Library work.

4.1.   Product Development

This is an example of how the tools included in the second release could be used to assist in the development of a Fortran 77 program. The example given here is of a single program and all its subprograms held in a single file. Large bodies of code would normally be held in multiple source files each of which could be processed separately. The Toolpack/1 tools referenced in this example include the editor, version controller, declaration standardiser, documentation generation aid, polisher, polish options editor, document formatter and macro processor. Detailed descriptions of these tools appear in the chapters of Part II. Rather than discuss the specific functionality of each of these tools here, our aim is to illustrate how a selection of Toolpack/1 tools can be combined to support the development of a software product.

(a) Development Cycle

During the development cycle the program undergoes a rapid series of changes while it moves towards completion. The Toolpack/1 editor is used to create and modify the program code as it allows certain common constructs to be entered as macros and prevents mistyping of keywords (a common cause of simple errors).

Once the code has been entered, it is used as initial input for a version control file, thereafter all changes are placed back in the version control file along with details of who made the change and why. To ensure that the programmer does not forget to update the version control file, the person responsible for installing the Toolpack/1 tools can ensure that the editor can be called only from a Toolpack script file which automatically recovers the latest version of the file, runs the editor, then invokes the Toolpack version controller to update the file.

The software is instrumented several times during its development to allow run-time trace information to be gathered and to ensure that the test data exercises as much of the code as possible. Cross-reference tables are generated using the documentation generation aid to assist in the development process. The declaration standardiser is used to ensure that all variables are explicitly typed and that definitions for unused variables are removed. The polish options are set to ensure that certain standards are adopted, in particular that all DO loops end on a unique label.

The documentation undergoes a similar development cycle with its own version control file. In order to simplify production of well presented documents, the Toolpack/1 document formatter and associated macros are used together to impose the chosen house style.

(b) Processing and Maintenance

Once the development phase is complete, the software becomes a product and is ready for release. There is a final phase of processing concerned with converting the code to a standard form, producing run-time statistics and filling in the mechanically generated sections of the documentation.

After the product has been released, the finished source and unformatted documentation can be used to establish new version control files. If the need arises to alter the source or documentation either to correct errors or incorporate enhancements, the relevant portions can be reprocessed using appropriate Toolpack/1 tools to ensure continuing conformance to the standards established during development.

4.2.   NAG Library Implementation

In this second example, we consider how the facilities of Toolpack/1 can be used in the implementation and maintenance of a large subprogram library, e.g., the NAG Fortran Library of numerical and statistical software. Present NAG implementation practice can be summarised as follows:

(1) a source text library and test software are assembled, standardised and tested on one system as the *base version*.

(2) the *base version* is then transported to *target systems* for adaptation and testing.

In the previous example (Section 4.1), the various steps taken were broadly applicable to NAG library activities (contribution, validation, standardisation, integration) prior to implementation. The emphasis in this second example is on transferring appropriate software to a new computing environment and certifying the behaviour of that software in the environment. The testing effort is directed towards that end so that the envisaged application of a wide range of static and dynamic analyses, as described in the first example, will probably not be repeated "en-masse" in the implementation phase. However, it is important that such facilities are available on the target system if required. The crucial difference between our circumstances, pre- and post- Toolpack, is the assumption that a set of uniformly designed, reliable tools will be available in all relevant environments. Indeed, we now expect that a preliminary activity of NAG personnel before embarking on active technical cooperation with a new project participant (contributor, implementor or otherwise) is to ensure that relevant parts of the tool suite are available on systems to be used by that person.

The steps involved in implementing a library of subprograms, assuming the availability of Toolpack/1 facilities, can be summarised as follows:

(1) The appropriate library subprograms with their associated test software, input data and reference results are assembled from master files.

(2) Using the Toolpack/1 Fortran-orientated editor, or some other transforming utility, anticipated systematic changes are made to the source code. The software so amended constitutes the *predicted version*.

(3) The amended version is transported to the target system for implementation.

(4) Depending on the effectiveness of stage 2, and the outcome of subsequent testing, further source code amendments may be necessary, these can again be made using the Fortran-orientated editor.

(5) Using the library compilation facilities of the target system, a compiled library is formed from the amended source code. The Toolpack/1 documentation generation aid can be used at this stage to derive callgraphs and other relevant information.

(6) The test software is executed using the compiled library. If errors are produced during the testing of particular subprograms Toolpack/1 analysis facilities may be used in determining the cause of the errors.

(7) Results produced by the execution of the test software are assessed against the supplied reference results. This may be done eventually by a data intelligent comparison tool. The Toolpack/1 text differencer may be used to compare the output with subsequent runs of the test in the same environment.

(8) Several tasks are involved in completing the exercise. These include preparing a library release tape and supporting user documentation. Unless step 4 involved no changes, it is advisable to perform a mechanical comparison of the source code with the original received version. This may be done using the Fortran aware comparison tool to avoid picking up comment and layout changes. The generated summary of changes will serve as a record of the textual divergence between the predicted and corrected versions. The summary in machine-readable form may be returned to the originating site for possible incorporation in the Toolpack/1 source code version controller. In producing supporting user documentation for the implemented library, the implementor can use the Toolpack document generation tools so that both formatted and unformatted versions can be returned.

We have described above the anticipated impact of Toolpack/1 facilities on a multi-machine library implementation exercise. The description could equally well be applied to other software packages intended to be mounted on a variety of systems. The Toolpack/1 facilities potentially provide a common functional processing basis on all systems involved.

5.   Conclusions

Release 2 facilities are not yet uniformly available across the many computer systems upon which NAG software is implemented but it is worth stressing that the tools in this latest release have been intensively used by NAG in assembling and processing Mark 12 of the NAG Fortran Library and other NAG products. This involves the processing of hundreds of thousands of lines of Fortran 77 source code within the context of NAG's stringent testing standards. Thus, whilst the public domain status of Toolpack/1 mentioned earlier should be borne in mind, NAG's own recent experience demonstrates that these tools can be effectively deployed in large scale software processing activities as well as meeting the needs of the individual programmer.

Structure of Toolpack/1 Software

Stephen J. Hague
NAG Ltd
Oxford, U.K.

*"A well-defined segmentation of the project effort
ensures system modularity."*

— Richard L. Gauthier and Stephen D. Ponto,
Designing Systems Programs, 1970

Abstract. The purpose of this lecture is to describe structural aspects of the
Toolpack/1 suite of software tools (both Releases 1 and 2). Topics covered include
the notions of tool fragmentation and integration, tool interfaces, the Toolpack Virtual
Machine, and the several tool operating regimes.

1.   Introduction

In the introductory Chapter "The Toolpack Project" we considered several examples
of software tools for processing Fortran 77 programs and associated items. One such
example was a "polish" tool which can be used to impose a consistent layout on
Fortran 77 software. From the user's point of view, this tool can be treated as a
source-to-source transformer, irrespective of the internal processing steps that might
be involved. Viewed in this way, the tool appears as a single *monolithic* utility. Indeed,
for much of the discussion, we could have been referring to a Fortran compiler. Such
compilers are usually also monolithic because they appear to users as large "black-
boxes" which transform their source code programs into a lower-level language form,
suitable for subsequent execution. Though the details may vary, these compilers
incorporate the same kind of components for lexical, syntactical and semantic analysis
as described earlier, but, instead of flattening and unscanning a program tree, they
enter a code generation phase in which the low-level instructions (variously called
relocatable binary, semi-compiled object modules) are produced. In general, the
internal components and data structures of the compiler are not accessible to the

23

A. A. Pollicini (ed.), Using Toolpack Software Tools, 23–32.
© *1989 ECSC, EEC, EAEC, Brussels and Luxembourg.*

user; a source program is presented and, if it is found syntactically and semantically valid, a compiled form of the program is returned.

The approach adopted by Toolpack is a distinct step away from the "black-box" philosophy which has been prevalent amongst most compiler writers. The Toolpack strategy has been to fragment tools wherever practical. *Fragmentation* involves identifying operations performed similarly in more than one major tool and then developing a separate tool to perform that operation, placing the output in a file. These "intermediate" files can then be accessed by a number of different tools rather than each having to create the required information independently.

The overall functionality of a user-level tool is achieved in Toolpack by combining a sequence of specific-function tools together. Although certain common processes (e.g. polishing) are provided as monolithic capabilities, these simply make use of the general fragments already available within the tool suite. Thus the overall effect of polishing may be performed by the discrete invocation of scan and polish fragments (in which case the token stream remains for further processing) or by invoking a monolithic polish function (in which case no intermediate token stream is output).

The sequence of tools required may also be given a name and presented to the user as if it were a "black-box" command, e.g.

"Polish" might mean: scan $\longrightarrow$ polish

"Apt"   might mean: scan $\longrightarrow$ parse $\longrightarrow$ apt $\longrightarrow$ polish

where "apt" is the name of an (actual) tool which converts the precision of a program in tree form, flattening to token form in so doing. If, in addition to applying "apt", we wish first to use tool "x" with similar input/output requirements to those of "apt", the tool sequence becomes:

scan $\longrightarrow$ parse $\longrightarrow$ x $\longrightarrow$ parse $\longrightarrow$ apt $\longrightarrow$ polish

or, in the case in which "apt" and "x" may be applied in reverse order:

scan $\longrightarrow$ parse $\longrightarrow$ apt $\longrightarrow$ parse $\longrightarrow$ x $\longrightarrow$ polish

i.e. it is not necessary to restore the program to source form and re-scan it before applying the second tool. It is true that the program must be reparsed between the use of "x" and "apt" but that has the benefit of validating the syntactical correctness of the program transformed by "x", and ensuring the associated symbolic/attribute information is updated. This example is illustrated in greater detail in the design overview of Section 2.1.

One of the benefits of fragmentation is therefore that unnecessary operations may be avoided. For instance, if it was required to generate further information from the previous example this could be done as shown in Figure 1.

Another advantage is that it facilitates the inclusion of additional tools (so long as they conform to Toolpack conventions), thus making the suite of tools extensible. It also permits considerable flexibility in the way in which the tools are presented to

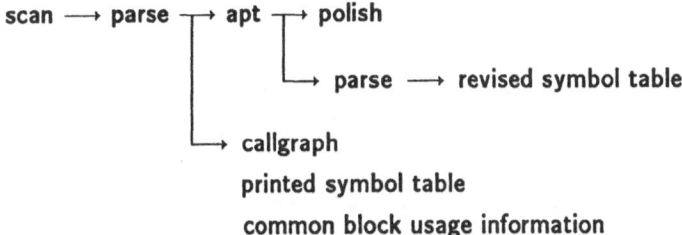

Figure 1. Tool Fragments as Building Blocks

users. The tools are "integrated" in the sense that they can be combined together in an appropriate sequence to satisfy a user's request for a specified result. The ways in which the tool sequence can be constructed and then invoked are described in the discussion of the user/Toolpack interface in Section 2.2.

The price paid for this fragmentation is the need to generate and use intermediate files. If no further processing of intermediate files is required then this is a costly and unnecessary step which can be avoided by combining the fragmented capabilities into single monolithic tools. Several such monolithic tools are provided in Release 2 of Toolpack/1. These versions always run quicker than their fragmented equivalents, but prevent additional processing steps being added to intermediate program forms.

2.   Design Overview

This section aims to give readers background information about the structural design of Toolpack/1, which consists of three categories of software:

(1) an integrated suite of tools;

(2) the user/Toolpack interface;

(3) the tool/system interface.

As discussed in Section 1, Toolpack/1 can be described as an *integrated*, extensible and portable suite of tools intended to support Fortran 77 programmers. The tool suite has an associated set of files (filestore); the tools process *objects* such as *program units* held in these files and place the results of their processing in other files. The term "Toolpack system" refers to the combination of the tool suite and the Toolpack filestore. We may also refer to this as an "environment" in that a developer may be able to undertake a succession of tasks using Toolpack/1 facilities only. However, Toolpack/1 does not provide a complete, or "closed", programming environment because major facilities (e.g. a Fortran 77 compiler) are not part of the distributed tool suite. On some host systems there will be a distinct separation between Toolpack tools and major host system facilities such as the compiler. For these "hostile" hosts a user needing access to a compiler or system text editor may have to leave the Toolpack/1 environment. On other systems however, this separation can be effectively hidden by mounting the tool suite so that the user can invoke host system facilities from within the Toolpack/1 environment. When mounted on these "friendly" host systems, Toolpack/1 can therefore offer a comprehensive and convenient "closed" support environment for the Fortran 77 programmer.

The user/Toolpack interface interprets user commands and invokes the appropriate sequence of tools. Each tool in the sequence expects to find its input in files created by the user or by other tools; in its turn, the tool writes its output to files. The tools call routines in the tool/system interface to perform input/output with the file store, including access to special files created by other tools. The tool/system interface also provides interaction between tools and the user/Toolpack interface as well as various "low-level" functions required by tools, such as character translation, bit manipulation, string processing, and time-stamping.

The user's view of Toolpack is presented by the user/Toolpack interface and the commands associated with particular tools. The user must know what results may be obtained and what commands must be issued to produce the results but he may use the tools effectively without knowing the details of their operation; moreover, the tool/system interface is invisible to him.

The three categories of Toolpack software are semi-independent in the sense that if any one is altered in certain prescribed ways to match the needs and resources of a particular site, the implied alterations of the other two components are well understood and easily made. The Toolpack installer is responsible for assembling the system from the available components and providing the necessary high-level documentation to users. The principal alterations permitted in each software category and their implications for users are as follows:

(1) The tool suite may be tailored to the needs of users – it may be enlarged to contain additional tools and may omit tools of no interest at a particular site. The set of available tools and their inter-dependencies are known to the user/Toolpack interface; hence tailoring of the tool suite implies that the user/Toolpack interface must be kept current.

(2) The user/Toolpack interface may be selected from amongst several now available but a site with the required programming resources and a strong interest in preserving a familiar command format may choose to develop its own interface. The choice of user/Toolpack interface reflects the level of sophistication in tool invocation and file management available to the user, as well as the cost of that sophistication in terms of systems resources.

(3) The definitions associated with the tool/system interface are the same for every host system but the software which implements these definitions may be in (reasonably) portable Fortran or may be tailored to a particular host, and need

not be in Fortran. Although the tool/system interface is invisible to users, its implementation strongly affects its efficiency in terms of execution time and the file space required for Toolpack. These definitions are described in [33].

Basic features of the software in each category are sketched out in the following three subsections.

2.1.   The Integrated Tool Suite

A summary of the major tools in the current suite may be found in Appendix A. The tools are written in Fortran 77. As mentioned above, each tool interacts with its environment through calls to routines in the tool/system interface. Whereas the tool writer must be aware of internal Toolpack/1 design conventions, none of these need concern the user. The important fact for the user is that the source code for the tools is identical – the strongest form of portability – in every host environment. Of course, different sites may select quite different user/Toolpack interfaces, so the user cannot expect to issue the same commands at a different site, but he can be sure that any differences are not in the tools themselves.

The following example illustrates the integration of the tool suite: a Fortran program in real single precision is to be transformed to double precision, the resulting program is to have its declarative part rebuilt and then the entire program is to be formatted according to user selected options. Precision transformation entails changes in declarations, representation of constants, format statements, and intrinsic functions.

The precision transformer tool does not expect all the declarations to be explicit. It is more sensible to apply precision transformation prior to declaration standardisation as the precision transformer will change the types of variables and may introduce new intrinsics. The declaration standardiser ensures that all the declarations are made explicit and are arranged in a systematic and uniform way with optional comments setting off the various types of declared names. Options enable the user to control the layout of the declarative section, to choose among legal possibilities for declaring arrays in common, and to remove the declarations of variables not used by the program.

Both the declaration standardiser and the precision transformer do their work by transforming a parse tree and symbol table to produce a revised token stream. The parse tree and symbol table are created from the token stream by the parser . The final token stream resulting from the transformations must then be reconstituted as Fortran source, which is done using the unscanner. The Fortran is formatted according to the preferences of the user as it is reconstituted from the token stream by the unscanner tool, familiarly known as "Polish" (unscanning with formatting options is sometimes called "polishing" a program.) The unscanner tool provides such a wealth of user options for formatting Fortran that an options editor is provided as a separate tool. The unscanner options are contained in a file read by the tool at run-time.

The sequence of actions to accomplish the specified transformations is shown in Figure 2. The name of each illustrated tool is given in square brackets, Toolpack name on the left and generic name on the right.

It should be noted that the sequence of tool operations ISTLX-ISTYP-ISTPT may be replaced by the "monolithic" tool ISTQT. Similarly ISTYP-ISTDS-ISTPL may be replaced by ISTDT.

In Section 2.2 we shall indicate ways in which the user may invoke the tools to perform such a sequence of operations.

2.2.   The User/Toolpack Interface

The interface that the user will see to Toolpack/1 depends on the way in which the tools are to be invoked. The interface is provided both by the tools themselves and the environment within which they are invoked. The following approaches may be adopted:

(1) Direct execution from the host operating system, using fragmented or monolithic versions of the tools.

(2) Execution from the host operating system by user provided *command files*.

(3) The Toolpack Command Executor (ISTCE); this may be used either directly or through the use of script files.

(4) The Toolpack Experimental Command Interpreter (ISTCI).

ISTCE and ISTCI may be regarded as tools – they are written in Fortran and conform to the tool/system interface. Their appearance to the user will be described in the Chapter "Toolpack Invocation Techniques". They utilise and give user access to the complete Toolpack filestore, which is described in detail later. The resulting operating regime is said to be *embedded*.

On the other hand, a user/Toolpack interface based on command files or direct tool invocation is host-system dependent. If it does not give access to the portable file store the operating regime is said to be *stand-alone*.

There are two additional operating regimes: *stand-atop* and *stand-astride*. A fuller explanation of the nature and uses of the operating regimes available is given in [33].

If the user/Toolpack interface insulates the user from any need to know dependencies amongst tools or how intermediate files are managed, the integrated system may be called a *programming environment*. The system uses existing files in a data base that it maintains, or invokes the necessary tools to create the files it needs but which do not exist in up-to-date forms.

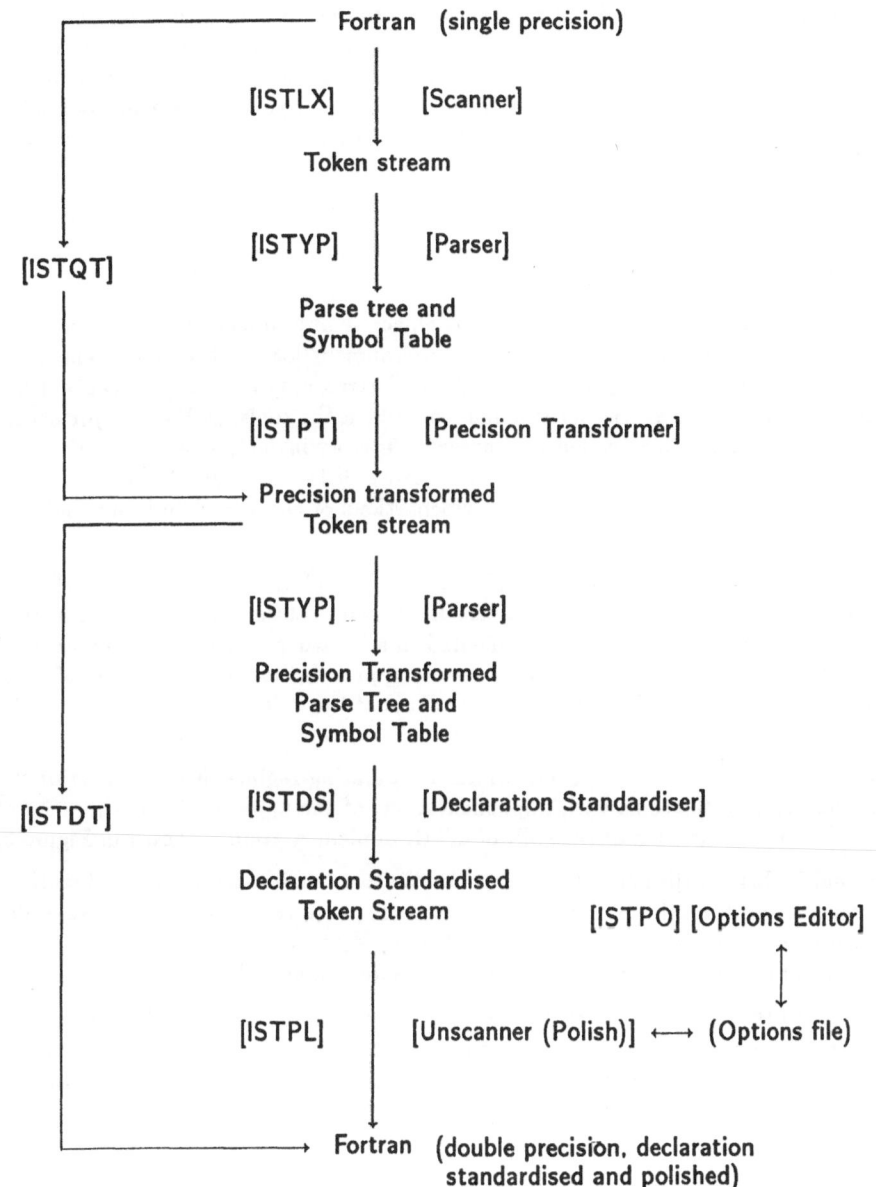

Figure 2. Tool Integration Example

The choice of user/Toolpack interface profoundly affects the way the user works in a Toolpack environment. The additional system resources required to support a more complex system may be offset by capabilities such as selective processing which reduces the tool invocations performed to a minimum. The choice for your site depends on the facilities offered by your host system and on the nature of your interest in Toolpack – whether primarily as a research prototype or primarily as a way of supporting existing program development activities.

2.3. The Tool/System Interface

A tool/system interface in the context of Toolpack is any implementation of a set of definitions called TIE (Tool Interface to the Environment)[33]. TIE defines a character set and a set of functions expressed as calls to Fortran subroutines. Conceptually, a tool/system interface is a "virtual machine" on which the tools, as Fortran programs, execute. TIE is sometimes called a "library". TIE is, in fact, a set of routine and capability definitions that define the characteristics of the *Toolpack Virtual Machine*. There can be, and indeed are, many implementations of the TIE definitions and it is these that really form the library.

Various implementations of TIE currently exist. One of them, called TIECODE, is provided entirely in Fortran 77 though certain routines contain host-sensitive Fortran constructions and a few must be implemented in a host-dependent way. Guidance on how to adapt host-sensitive code and on how to approach the implementation of host-dependent primitives may be found in Appendix B of the *"TIE Installers' Guide"* [30]. An implementation tailored to a host system will be more efficient.

TIE is divided into sub-libraries. The different operating regimes listed in Section 2.2 result from loading the tools with implementations of the appropriate sub-libraries of TIE. A conceptual diagram of the split of a TIE implementation is shown in Figure 3.

Additional facilities, called *supplementary libraries*, provide important special functions such as accessing the information in token stream, parse tree, and symbol table files produced by the scanner and the parser. Other supplementary libraries offer extended string and table handling facilities and portable screen manipulation facilities.

The existing TIE definitions are stable, however a redesign of the portability base in the light of installation and tool writing experience may be undertaken at some time in the future. Supplementary libraries may change at each release and additional libraries, e.g. for graphics, may be defined and implemented.

3. Conclusions

The flexible, extensible design adopted by the Toolpack project, and reflected in the first two releases of Toolpack/1, present the tool installer and the tool user with a number

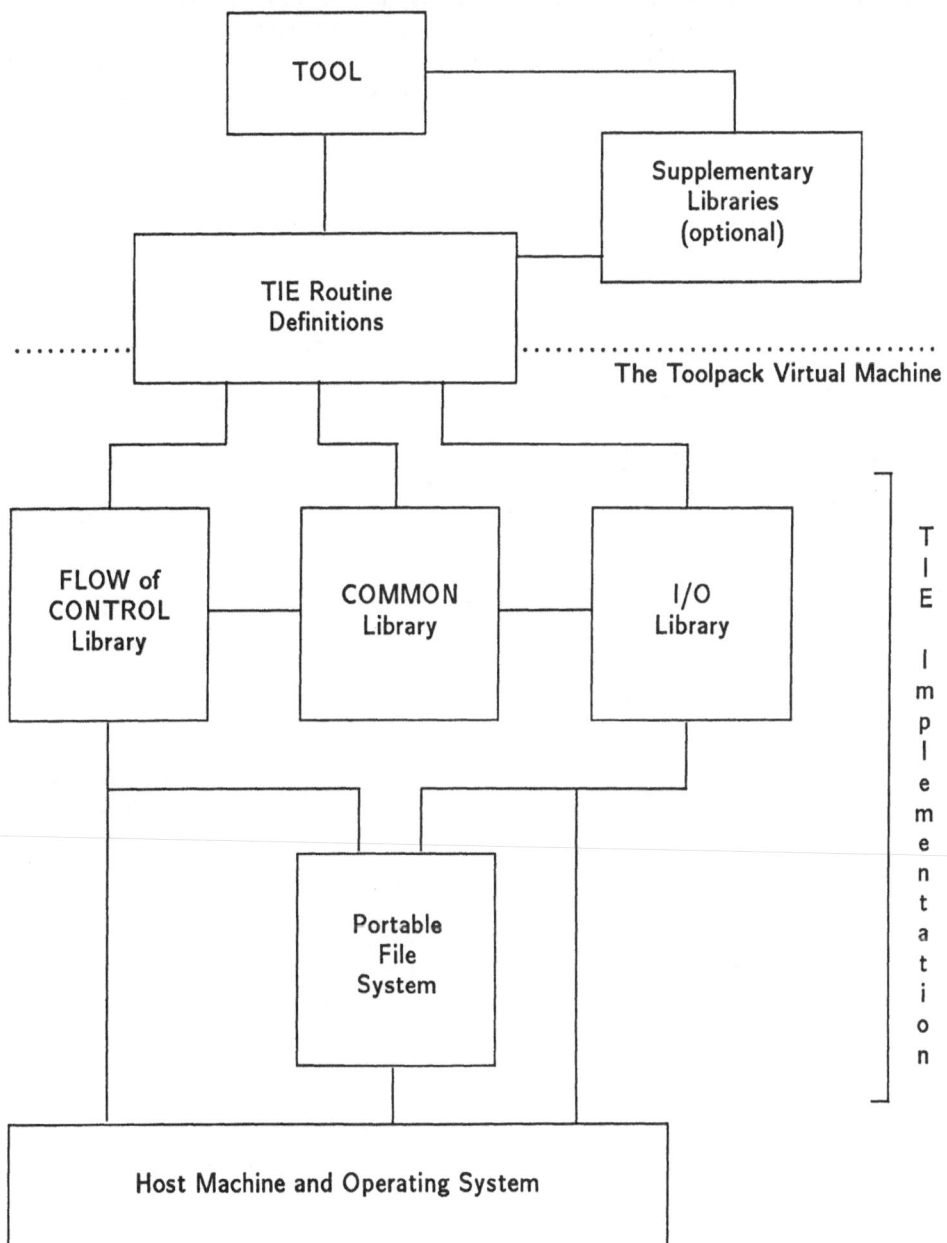

Figure 3. Diagram of Tool/System Interface

of options. Presenting these options in a clear way is a major challenge to be overcome if Toolpack/1 facilties are to be widely adopted. Given the aims of the project and the diverse, conservative nature of the Fortran user community, the Toolpack designers felt that it was better to confront this challenge rather than opt for a single rigid design.

TIE Specifications

Malcolm J. Cohen
NAG Ltd
Oxford, U.K.

> *"Our most important mental tool for coping with complexity is abstraction."*
>
> — Niklaus Wirth,
> On the Composition of Well-Structured Programs,
> Computing Surveys 6, 4, 1974

Abstract. The specifications of the Tool Interface to the Environment are presented. The way in which their implementation as TIE library provides the portability base for Toolpack/1 software tools is also discussed.

1.   Introduction

This chapter describes the evolution and specification of the Toolpack/1 portability base. These specifications describe the Toolpack Virtual Machine (TVM), and are known as the Tool Interface to the Environment (TIE). These specifications were developed from the earlier Fortran I/O library FIO (described in [46]), and strongly influenced by the Software Tools library of Kernighan and Plauger (described in [41]). FIO was a development of the Portable Directory System of Hanson, which is described in [26].

2.   Overview

To provide the capabilities of the Toolpack tools and command interpreters on as wide a range of machines as possible, a definition of the Toolpack Virtual Machine (TVM) was produced to which all Toolpack programs comply. The TVM includes definitions of the character set, directory structure and minimum machine capabilities that must

A. A. Pollicini (ed.), Using Toolpack Software Tools, 33–43.
© 1989 ECSC, EEC, EAEC, Brussels and Luxembourg.

be provided on the host computer before Toolpack programs can be used. The TVM capabilities are actually accessed by the tools via the Tool Interface to the Environment (TIE) library.

In order to make the capabilities of the TVM available, it is necessary to have an implementation of the TIE library available on the host machine. A relatively portable version of the TIE library, written in Fortran 77, is provided in the Toolpack/1 software suite to allow as many users as possible to at least try out Toolpack capabilities. The needs of portability have, however, led to this "vanilla" implementation of TIE being rather inefficient. There are now a growing range of machine-tailored TIE implementations being made available that greatly improve the efficiency of Toolpack tools.

The main features of the TVM are as follows:

(1) A stream-based I/O system for files and pre-connected units.
(2) A fixed internal character set and variable length string handling.
(3) A tree-based directory structure and defined file naming convention in a portable file store (PFS) normally separate from the host file store (HFS).
(4) A defined process scheduling and argument passing capability.
(5) Extensible capabilities by the use of supplementary libraries.

By splitting the TIE library into several sections it has been possible to allow greater environmental flexibility for the use of tools. This division is shown in Figure 3 of the Chapter "Structure of Toolpack/1 Software". This shows that there are three sub-libraries to TIE: **common, input/output** and **flow-of-control**.

The common sub-library contains general purpose routines to convert character types, manipulate strings and recover the date and time. There is only ever one version of this library in a TIE implementation.

The input/output sub-library is concerned with the provision of the stream-based input/output routines and the directory handling capabilities of the portable file store.

There may be two separate implementations of this sub-library, one of which provides the full capabilities of the host and PFS file stores as defined for the TVM and the other which maps all input/output and file access to host files and ignores directory handling routines. The two versions of the sub-library contain the same routines with the same apparent (to the tool) functionality. Therefore, no code changes are required to use either version.

The flow-of-control sub-library is concerned with the initialisation and termination of Toolpack tools, process scheduling and argument passage. There may also be two versions of this sub-library, one which allows process scheduling and the other which does not.

3. Evolution

3.1. The Portable Directory System PDS

The Portable Directory System (PDS) was developed by Hanson in 1979. It was a set of subroutines and functions which provided a machine-independent directory system which was used to access ordinary host files. This operated simply by providing a mapping from "PDS" names to host names.

The directory structure provided by the PDS was essentially the same as that of Unix. That is, a complete filename (called a *pathname*) specifies a list of directory names ending in an ordinary filename; the components of the pathname are separated by slashes. A pathname beginning with a slash specifies a path starting from the top-level, or *root* directory; with no slash the path specified begins in the current working directory (default directory).

3.2. The Fortran I/O Library FIO

The FIO library was an extension to Hanson's PDS scheme which included not only a directory system but also an I/O system and character set. I/O to files in the portable I/O system was performed through FIO subroutine and function calls, which accepted a portable character set (comprised of small integers) and buffered all input and output internally. The files in the portable I/O system were stored in "device" files which were large direct-access files in the host system. I/O to host files continued to be done using Fortran READ/WRITE statements.

The development of TIE from FIO was driven partly by the need to have a stable and consistent set of routines providing host-dependent operations. An additional consideration was conformance, where practical, with the routines specified in Kernighan and Plauger's *"Software Tools"* [41]. The reason for this was to ease the adoption of existing software tools which had been written to conform to the latter definitions.

4. Specification

TIE stands for the Tool Interface to the Environment, a stable, portable and extensible definition of the Toolpack Virtual Machine (TVM). The functional capabilities of the TVM are defined by specifying the inputs, outputs and operations of routines that would implement the functionality. TIE and any software implementation of TIE are always considered separately, thus allowing development to take place on one or more implementations without compromising the stability and portability of the TIE definition. The definition of a routine is guaranteed to remain stable once accepted into TIE, so it may be used with confidence by both TIE implementors and tool writers.

TIE is not intended as a replacement for the capabilities of the host operating system but as a controlled Fortran 77 *microclimate* within which tools can be generated, maintained and used portably and easily. TIE does not attempt to provide all possible required routines, more a kernel of routines on which portable tools can be built. The major use of TIE is in the support of the **Toolpack** software that provides the **Integration System for Tools (IST)** although it may also be used for the production of portable application software.

The complete definition of TIE is contained in *"The TIE Routine Definitions"* [35].

4.1.   The Toolpack Virtual Machine

The Toolpack Virtual Machine (TVM) is a pseudo Fortran 77 operating environment whose capabilities are defined within TIE and provide the following:

(1) A library of commonly used routines to perform a number of general and specific operations within the virtual machine.

(2) Two separate file stores, one general purpose with limited capabilities (host) and a Portable File Store (PFS) offering much greater functionality as well as a portable file naming and directory structure.

(3) Two character sets and string types: strings of pre-defined length as defined in the Fortran 77 standard [1] using the host Fortran 77 character set, and a portable, variable length string structure using an internally defined (and therefore portable) character set. This second form is called *IST Strings*.

(4) Pre-connected devices for input, output, error logging and a hard copy list channel.

(5) A unified input/output (I/O) system for communication using either character set to either file store or to the pre-connected units.

(6) A set of four operating regimes for tools and application programs.

(7) A set of optional extensions for handling a variety of pre-defined entities (e.g. parse trees, token streams, VT100 compatible terminals etc).

The manner in which these capabilities are implemented on any given host machine may be important in operational practice, but the Toolpack Virtual Machine is concerned only with their definition. As long as the TIE definition is implemented accurately the tools will be unaware of the host machines' operation. Note that no definition of the TVM "hardware"-is provided. TIE provides means of packing and unpacking bytes and characters and of discovering information about the integer and character sizes of the host, should these be required.

The nature of the capabilities is expanded in the following subsections. The terms "tool" and "tool writer" may be taken as including "application program" and "application program writer". The term "TIE implementor" refers to any person or group that produces an implementation of TIE. The term "TIE installer" refers to any person or group that installs an existing TIE implementation (e.g. TIECODE, TIEVMS) onto a host machine.

The following subsections summarize briefly the characteristics of the TVM.

4.2.   TVM Functions

The TIE library is split into three logically distinct areas; Common, Flow-of-control and I/O.

The functions specified in the Common section provide various general capabilities. Routines are defined to supply the current time and host specific values, to carry out logical operations and character conversions, and to manipulate IST strings.

The routines defined in the I/O section provide access to the TIE unified I/O system for handling and modifying files, which is described below. Routines are defined for character and line input and output, integer output, error message output, and file creation, opening, positioning, closing and deleting.

The routines in the Flow-of-control section reflect the fact that Toolpack tools may be run as sub-processes to a command interpreter. Routines are defined to enable the tool writer to create a command interpreter (i.e. a tool which schedules other tools) or a tool (i.e. a scheduled tool). The method of scheduling (spawning, chaining, etc.) is not defined; either is possible in an implementation. Routines are also defined for the passing of status, arguments and results between a tool and the command interpreter.

4.3.   The Flow-of-Control Sub-library

The major routines in the flow-of-control sub-library are:

ZINIT   The TIE initialisation routine; this routine must be the first TIE routine called by any TIE-conforming tool.

ZQUIT   The TIE termination routine; this routine terminates the tool with a specified status. All tools must terminate either using this routine or the routine "ERROR".

GETARG   This function returns the arguments to the tool. These may be either command-line arguments (picked up by the TIE implementation) or arguments passed from a TIE-conforming command interpreter.

The remaining routines in the flow-of-control sub-library provide facilities purely for the use of command interpreters (e.g. tool invocation and argument passage) and so will not be dealt with in this chapter.

4.4.   The File Stores

The TVM provides the capability of accessing either the host file store (HFS) or the Portable File Store (PFS). Both file stores can handle character sequential files or "direct access" files containing fixed length IST strings.

Access to the host file is limited to ensure that it can be made available, in some form, on as many machines as possible. Files may be created, destroyed, opened, closed and tested for existence. The contents of sequential files can be read on a character basis, written on a character basis or positioned on a character basis. The contents of direct access files can be read and written on a record basis. File names for the host store are host-system dependent and are always preceded by the *hostfile_id* character (*sharp* or *hash*). The system manipulates "normal" host files that may be accessed outside a tool using the standard host operating system utilities.

The PFS is a more fully defined and controllable file store that provides all the capabilities of the host store plus a rooted tree directory structure. The directory structure may be manipulated using TVM functions. The PFS recognises file names of a fixed form regardless of the host file name type. A file in the PFS has a pathname that fully defines the location of the file with respect to the root of the system. The pathname starts with a '/' (the root) and consists of a list (possibly empty) of nested directories separated by '/'s and terminating in a file name. Individual file and directory names are held as IST strings, may be up to 12 characters long (plus an *eos*) and may include '.' and '-', letters, and digits but must start with a letter. The full pathname is limited to 80 characters (plus an *eos*). Character case is ignored in the pathname. If no preceding '/' is provided on the pathname then a search for the file starts in the current local directory. The local directory may be changed using TVM functions.

The full capabilities of the PFS file store may either be implemented directly on the host (if the required capabilities exist) or by a portable directory and I/O system such as that provided in the TIECODE implementation. Depending on the implementation, files within the PFS may or may not be directly accessible to the standard host operating system and utilities. They can, of course, be copied to host files using the Toolpack/1 command executor, ISTCE.

4.5.   The Character Sets

Two character sets (and strings of these characters) are recognized by the TVM, namely Fortran 77 characters and IST characters.

Fortran 77 characters and strings may be manipulated as defined by the Fortran 77 standard. TIE defines input and output routines for Fortran 77 characters and strings and provides for conversion between them and IST strings. Variable length strings may be used for output as the '.' character is taken as a string terminator, a '.' may be output by placing it twice in the string (i.e. '..').

IST characters are small integer values held in integer variables, one character per variable. IST strings are integer arrays containing one IST character per element and terminated by a fixed end-of-string character. The end-of-string character, known as *eos*, is an additional value which differs from all "real" characters. This definition allows for easy handling of variable length strings. The numeric values of the IST characters are defined within TIE, thus making the character set "portable" across implementations. All mapping between IST and host characters can be performed using TIE routines and is performed automatically during I/O. Use of a mixture of the two character sets on a single I/O stream is possible.

Additional capabilities are provided within TIE for handling Hollerith constants and converting between them and the IST character set.

4.6. Pre-connected Devices

Four pre-connected devices are recognized: standard input, standard output, a standard error log, and a standard list unit. These units are always available to a tool and do not require opening. Input-output streams may be opened to the pre-connected devices by using any of their reserved file names:

```
"0" or "#0"      standard input device
"1" or "#1"      standard output device
"2" or "#2"      standard error log device
"3" or "#3"      standard list device
```

The standard list device is normally a hard-copy printer but its implementation is host dependent (it may be a normal disk file which the user can print after tool execution). Files connected to this channel may be flushed to the printer using the TIE routine ZSPOOL.

4.7. The Unified I/O System

The unified I/O system is concerned with providing the means of communicating with the pre-connected units and both file stores using either character set. The capabilities provided are appropriate to the file types defined; sequential and direct access. Up to 16 files (sequential, pre-connected and direct access) may be opened at any time, four streams are reserved for the pre-connected devices and the remaining streams may be used for access to files in either of the file store systems.

TIE defines all sequential files as being formatted streams. All stream output is buffered internally except error messages using ERROR or REMARK and user prompts using ZPRMPT. Operations defined for the stream package are as follows:

input Characters may be read from an input stream using get-character or get-line routines. Lines are defined as variable length character strings terminated by a *newline*. Input may be in either character set. Input represented in the IST character set has a trailing end-of-string marker.

output Characters may be put in an output stream by using put-character or put-line routines. Lines have variable length and are terminated by a *newline*. All output is buffered; the buffers are flushed when a *newline* character is output, the stream is closed or the line length is exceeded. Output routines are defined for integers and for both character sets.

position A stream may be positioned to any character. Character positions may be defined in absolute terms or as positive relative offsets. Pre-connected units may not be positioned.

open A stream may be opened for access to a file if the file already exists in either file store. The selection of the file store depends on the name of the file given as an argument in the OPEN routine. The stream may be opened for read, write or read/write access unless it is a pre-connected device. The file is rewound after being opened.

create A stream may be created for access to a file whether the file already exists or not. If the file does not exist it is created and opened. If it does exist it is opened and truncated (i.e. the existing contents of the file are lost). Again, the selection of the file store is made by the choice of file name.

close A stream may be closed at any time. All streams are closed automatically when a tool terminates.

remove A file may be removed or deleted from the host system. (The effect is the same as a Fortran CLOSE with STATUS = 'DELETE')

Direct access files are provided that handle records one line in length. Direct access uses the same close and remove utilities as in sequential access but special create, open, read and write utilities are provided as follows:

open A TVM utility opens an existing direct access file and returns the maximum record number for the file.

create A direct access file with a given name and a specified number of records may be created. Creating a file with a given name overwrites any file of that name existing at the time.

input The ability to get any record from any opened direct access file is provided. Records run from 1 to the number of records in the file.

output A record may be written to any open direct access file.

4.8.   The Tool Operating Regimes

Four operating regimes within which tools may be executed are defined in TIE. These regimes are called **embedded, stand-astride, stand-atop,** and **stand-alone.** The principal regime for Toolpack is embedded, though tools are often developed stand-atop; the other regimes may set limits on the type of operations valid on the file stores.

In some implementations of TIE the differences between these regimes may be less pronounced than detailed here, or even non-existent.

In the embedded regime, tools are scheduled as sub-processes to a central command interpreter or executor. When the tool terminates, control returns to the command interpreter. Both file stores are fully available. In this regime the command interpreter may make extensive use of the PFS, setting up directories and files that it requires for its operation. These special files are "off limits" to tools that have no knowledge of their purpose or structure.

In the stand-astride regime the same flow-of-control as provided in the embedded regime is available. Access to the PFS is not possible (indeed it need not even exist) and all PFS file actions are automatically mapped onto host files. PFS actions that fall outside the capabilities of the host system will be ignored.

In the stand-atop regime both file stores are available, as in the embedded regime. However, the flow-of-control is different in that tools are executed by the host operating system and, on termination of the tool, control returns to the host operating system.

In the stand-alone regime flow-of-control is as in the stand-atop regime but only the host file store is available, as in the stand-astride regime.

Tools are normally written to run in any of these regimes without prior knowledge of which environment will be used, the selection of the environment is normally done during loading of the tool when it is necessary only to select a different instance of the local TIE implementation (i.e. no code changes in the tool are required).

4.9.   The Common Sub-library

The common sub-library provides the following capabilities:

(1) Conversion between integers and IST strings (decimal): CTOI, ITOC, ZITOCP.

(2) Conversion between IST characters, Fortran 77 characters and Hollerith characters: ZCCTOI, ZCHTOI, ZCITOC.

(3) Conversion between upper and lower case: ZLOWER and ZUPPER.

(4) Conversion from a PFS time stamp to an IST string: ZTIMES.

(5) Comparison between two IST strings: EQUAL, ZCOMPR, ZORDER.

(6) data describing the host machine (bits per integer etc.): ZHOST.

(7) the current date and time: ZTIME.

(8) the name of the TIE implementation: ZIMPLS.

(9) byte operations: ZBYTE, ZCBYTE.

(10) bit field extraction: ZFIELD.

(11) logical operations: ZIAND, ZINOT, ZIOR, ZLLS, ZLRS.

(12) examining an IST string: ALLDIG, INDEXX, LENGTH, TYPE, XINDEX, ZINDEX.

(13) extraction from an IST string: GETWRD, ZSBSTR.

(14) manipulating IST strings: ADDSET, SCOPY, SET, SKIPBL.

5.   Optional Extensions

TIE allows for extension of the virtual machine by the use of "supplementary libraries", each covering some application area. The capabilities defined in these libraries are not required of an implementation of the virtual machine so that the underlying definition of TVM remains stable. In general, supplementary libraries are mutually independent although each depends on the facilities of TVM.

Some widely used tools may depend on the existence of supplementary libraries. Further, as the repertoire of tools grows, tool writers may wish to extend the TVM performance by providing additional supplementary libraries. Such libraries may then be distributed and come into common use but it is important that they be regarded as distinct from the basic definition of the TVM.

Existing supplementary libraries are:

ACCESS   provides routines for accessing and manipulating the defined alternative representations of Fortran 77 programs. These representations are: token stream, parse tree (with symbol table), attributed parse tree (with extended symbol table) and flowgraph. The token stream and parse tree representations are described in the Chapter "Analysis Tools for Fortran 77".

STRING   provides additional routines for creating and manipulating strings (both IST and CHARACTER). These routines include pattern matching and replacement primitives.

TABLES   provides routines for manipulating ring buffers, sparse arrays, tables, binary trees, lists and heaps.

WINDOW   provides routines for manipulating terminal screens. This library is more host-dependent than the others and so less likely to be available at a particular site.

6. Implementations

6.1. The TIECODE Implementation

A reasonably portable implementation of the TIE definition has been produced as a starting point for all sites interested in Toolpack. This implementation is known as TIECODE and has been installed under a number of machine/operating system combinations. TIECODE achieves an implementation of the PFS by providing modified versions of a portable directory and I/O structure originally designed by Hanson. This system requires that two large "device files" are made available in the host system to support the data and directory structures of the PFS.

6.2. Host-Specific Implementations

There are currently 2 other full implementations of TIE provided with the distributed software. One is for VAX/VMS systems and is written partly in Pascal. The other is for Unix systems and is written in C.

Toolpack Invocation Techniques

Stephen J. Hague
NAG Ltd
Oxford, U.K.

"What's in a name? that which we call a rose,
By any other name would smell as sweet;"

— William Shakespeare,
Romeo and Juliet, act II, sc. 2

Abstract. This chapter deals with the user interface to the Toolpack suite of software tools. It describes the several ways in which users may call (*invoke*) those tools.

1.   Introduction

Even though a software package should possess a number of desirable properties such as reliability or efficiency, it is now generally appreciated that the quality of its user interface is a very important factor in determining its fate in the marketplace. Indeed, in the present day world of personal computers, colour graphics and WIMPs (Windows, Icons, Menus and Pointers), some would argue that the user interface is more important in the marketplace than the quality (or otherwise) of the underlying software. Any software tools project launched in the mid 1980s would no doubt pay close attention to presentational aspects as well as to questions of functionality or internal software design. Toolpack was conceived in the late 1970s, however, and whilst its designers recognised the importance of the user interface, a number of factors influenced their thinking and thus affected the eventual interfaces developed during the course of the project. At the time of Toolpack's inception, the development of portable, screen-oriented interface facilities (to say nothing of graphics) was considered beyond the scope of the project. Thus, the forms of user interface of Toolpack reflected the project's pre-occupation with portability and flexibility, rather than a drive towards a WIMP-style interface.

In the following sections we describe four forms of user interface available for **users** of Toolpack tools. Because of the modularity and flexibility of Toolpack **design**,

45

A. A. Pollicini (ed.), Using Toolpack Software Tools, 45–52.

other forms could be provided, and indeed alternatives or enhanced forms have been developed in the recent past. Most notably, a form of interface suitable for modern scientific workstations has been developed at the University of Kent at Canterbury [5]. Further developments of this kind may be expected in due course.

2.   Toolpack User Interfaces

The four interfaces described below are as follows.

(a) Direct Invocation;

(b) Host-dependent command files;

(c) ISTCE – Command line executor;

(d) ISTCI – Object-oriented interpreter.

The descriptions given assume familiarity with the Toolpack filestore terms, PFS and HFS, which are introduced in the Chapter "Structure of Toolpack/1 Software".

2.1.   Direct Invocation

The simplest form of tool usage to set up is direct invocation. In this approach, tools are treated as ordinary host system programs that may be run individually. The naming and saving of intermediate files, and the correct sequencing of tool invocations, are the responsibility of the user. This approach is sufficient if the monolithic tools are available, or if the users feel happy with invoking low-level tool functions.

Toolpack tools receive their arguments in a TIE implementation dependent manner, usually from a named host file. The use of a host file allows the 10 arguments (each potentially a line long) and other information (e.g. current local directory) to be passed easily, but it is not very convenient for the user. To enable a tool to pick up its arguments in this way would require the user to create a mock file for the tool to read. Because of this, all tools will prompt for arguments that are not supplied, so it is only necessary to run the tools and then supply information in response to the prompts. Some TIE implementations actually use the normal host argument passage techniques, this allows arguments to be specified at invocation.

When using direct invocation the monolithic forms of the tools are often more convenient as intermediate objects, and multiple tool invocations may be avoided. The only disadvantage of using the monolithic tools is that only some functions are provided (i.e. tools such as ISTME, a tool for manipulating arithmetic expressions, do not have monolithic versions) and additional processing on intermediate files is not possible.

2.2. Host-Dependent Command Files

The second approach is to isolate the user from the need to know about Toolpack intermediate files when using fragmented tools. This can be done through the use of host script or command files. This approach is particularly useful where a lot of software is to be processed by a fixed tool sequence or where only a limited number of different forms of processing are to be supplied. Both monolithic and fragmented tools may be invoked from command files.

As an example, Unix command files (shell scripts) have been written to manage tool invocation and file management for a subset of the Toolpack tools. These scripts offer neither the portability of ISTCE nor the generality of ISTCI but they provide a simple Fortran tools environment in the context of Unix. Argument passage is effected by the provision of a special program for Unix sites.

To illustrate the use of such scripts, Fortran source may be instrumented for dynamic analysis (assertion testing, execution frequencies by program segment, and execution tracing by program segment) by the Unix command line:

```
inst <Fortran source> <instrumented Fortran source> [save]
```

where <Fortran source> is the name of a file containing one or more Fortran program units that may contain embedded assertions. The instrumented source is placed in <instrumented Fortran source> and additional files are created for use by a script that reports the results of running the instrumented Fortran. Output (sent to standard output) is a report consisting of two parts:

(1) A program listing annotated with segment and assertion identification numbers;

(2) A static count of Fortran keywords and other program elements.

This script invokes ISTLX to produce a token stream, ISTAN to instrument the Fortran source, and ISTAL to produce the output report. If the third argument is set to "save", the token stream produced by the lexer will be saved for use by other scripts.

As a second example, the precision transformation, declaration standardisation, and program formatting depicted in Figure 2 of the Chapter "Structure of Toolpack/1 Software" may be accomplished by the following command line:

```
dapt <s or d> <decs options> <option file> <Fortran source file> [save]
```

If the first argument is 's' (or 'd') the Fortran in <Fortran source file> will be transformed to a REAL (or DOUBLE PRECISION) version. The next argument is the name of a file containing declaration standardisation options. "Option file" is the name of a polish option file that may be created using the Toolpack Polish options editor invoked, in turn, by a script called polx. The optional fifth argument permits saving the output from the Toolpack lexer and parser.

2.3.   ISTCE

The portable command executor ISTCE provides a means of maintaining the PFS directory structure in an embedded environment and may be used to perform single or multiple tool invocations. Tutorial examples of the use of tools via ISTCE are given in Appendix A of the *"Toolpack/1 Release 2 Introductory Guide"*[17].

Each tool has a two character identifier (e.g. LX identifies the lexer/scanner named ISTLX). A tool may be executed by entering its identifier as a command; the user will be prompted for the arguments. Alternatively, some or all of the arguments may be specified on the ISTCE command line. Arguments are separated from the tool name by at least one space and from each other by commas. Commas can be used as space holders for arguments to be input in response to a prompt. ISTCE also provides a set of basic commands to assist in maintaining the Toolpack enviroment. These are summarised below ([...] denotes an optional argument, ... | ... denotes permitted alternatives):

```
#                                   COMMENT LINE
?                                   LIST COMMANDS
!                                   LIST TOOLS
CD <Directory name>                 CREATE DIRECTORY
CF <File name>                      CREATE FILE
CL                                  CLEAR FLAG
CP <File name>,<File name>          COPY FILE
DD <Directory name>                 DELETE DIRECTORY
DF <File name>                      DELETE FILE(s)
EC [<+|Y|y|-|N|n>|"string"]         ECHO
EX                                  QUIT
LD <File name>                      DISPLAY DIRECTORY
MV <File name>,<File name>          MOVE FILE
RN <File name>,<File name>          RENAME FILE
SC <File name>                      USE SCRIPT FILE
SH <File name>                      SHOW FILE
SK [<count>]                        SKIP IF NOT OK
TF <File name>                      TOUCH FILE(S)
TM                                  TIME
WD [<name>]                         WORKING DIRECTORY
XS <File name>                      CHECKSUM
```

We now give two examples of tool use in an embedded environment.

The command

```
LX #SRC,1,LX.TKN,LX.CMT
```

would invoke the lexer on the Fortran source in HFS file 'SRC', sending the token stream and comment stream to PFS files 'LX.TKN' and 'LX.CMT' respectively. Error messages and a listing annotated with the number of the first token in each statement would be sent to standard output (pseudo-filename '1').

The command

```
PL LX.TKN,LX.CMT,,#PL.OPT
```

would invoke the Fortran program formatter on the token and comment streams produced above by the lexer. The user would be prompted for the name of the output file (third argument) as it was not specified. The formatting would be governed by the options in HFS file 'PL.OPT'. If an argument needs to include a comma (or leading or trailing spaces) then it should be enclosed in double quotes "like this"; a double quote may be entered by typing it twice, "".

Command lines may also make use of the 9 local variables &1 to &9 available within ISTCE. Each of these variables can be set to a string using the commands S1 to S9 and these strings can then be recovered using &1 to &9. Substitution of local arguments takes place prior to interpretation of the command; for example, the command sequence

```
LX SRC,1,LX.TKN,LX.CMT
PL LX.TKN,LX.CMT,POL,PL.OPT
```

is the same as

```
S1 LX
LX SRC,1,&1.TKN,&1.CMT
PL &1.TKN,&1.CMT,POL,PL.OPT
```

ISTCE is also capable of executing script files stored in the directory '/COM' or elsewhere. Any command that is not recognised by ISTCE as being either an internal command or a tool scheduling request is assumed to be the name of a script file in the directory '/COM'. Script files may take up to 9 arguments on the command line, interpreted in the script as the local variables &1 to &9 (the variable &0 is the name of the script file itself). Script files may be nested up to 10 deep. A script file not in '/COM' may be executed using the SC command.

To simplify the writing of script files, and to make them reasonably portable between Toolpack sites that use the embedded environment, a "preferred" method for storing and naming files within the PFS has been devised. This is not a compulsory method, just an approach that has been found to make the use of script files easier. It is recommended that a directory '/PU' (PU is for program unit) is set up to hold Fortran 77 files. Each directory under '/PU' is associated with one Fortran 77 program unit. The program unit is in a file called 'SRC' under the appropriately named directory. For example, the contents of the host file 'FRED.FTN' (containing the program unit FRED) would be stored in the PFS file '/pu/fred/src'. All the other files derived from

FRED are also stored in this directory. A list of the "preferred" names is included in Appendix B of the *"Toolpack/1 Release 2 Introductory Guide"*[17]. The following is a script file called POL that makes use of this fixed naming convention to polish (unscan and format) a Fortran 77 file:

```
WD /PU/&1
LX SRC, -, LX.TKN, LX.CMT
PL LX.TKN, LX.CMT, POL, PL.OPT, &2
WD /
```

The script file can be invoked by simply typing

```
POL FRED
```

from within any directory. The 'WD' command changes the current local directory to '/PU/FRED'. This is followed by the lexer command 'LX' which is informed that the source code is in a local file called 'SRC', no list file is required. The third command, 'PL', invokes the polish tool. Its final argument, '&2' allows a single option override for the unscanner to be specified, e.g.

```
POL FRED, LMARGS=10
```

The script file POL changes the local directory to the name specified, then uses the "preferred" names as it invokes the scanner and unscanner tools. On completion, the local directory is set to the root.

Because command lines are expanded before being interpreted it is also possible to set up script files that use arguments in place of commands; thus the following script file (MYPROG) will perform the operation named on the command line on the three files FRED, TOM, and BILL.

```
&1 FRED
&1 TOM
&1 BILL
```

To polish all of the files mentioned in MYPROG the following command can be entered:

```
MYPROG POL
```

A number of useful files for commonly occuring tasks is provided with ISTCE. These files, described in further detail in Appendix A of the *"Toolpack/1 Command Executor"* [29], are listed below:

POLISH <file>

> The POLISH script file will replace a source file with a polished version of itself. If any errors occur during the scanning or polishing operations then the original

source is not replaced and the intermediate files (including the scanner list and error files) are not deleted.

PRINT <file>

This script will process the document in '/pu/<file>' if one exists. The result of the process is always a file called TXT. If a file called ANL exists then this is processed to DOC, and then to TXT. If a file called DOC exists then this is processed to TXT. In the event of a file called ANL existing, it is up to the user to ensure that any objects that will be required by ISTAL are available and up-to-date.

IMPORT <dir>,<file>

This script file will set up a directory called 'pu/<dir>' and will store the contents of <file> in 'pu/<dir>/src'. This operation sets up a source file in the "preferred" manner for processing by the other script files provided.

PROCESS <file>,[<option>],[<polish-option-override>]

The PROCESS script file will perform any one of a number of operations on the specified file. The options accepted are 'GI', 'DS' and 'PT' to perform replacement of references to specific names of intrinsics by their generic forms, declaration standardisation or precision conversion respectively. If no option is specified then a polish operation is performed. If no errors occur, each operation ends in a polish (using 'pl.opt' with the optional argument override) and produces the files 'LX.POL', 'GI.POL', 'DS.POL' and 'PT.POL'. If any of the intermediates required by the specified operation already exist and are up-to-date, then that stage of the processing is skipped. Precision conversion is always followed by declaration standardisation, and the replacement of references to specific names of intrinsics by their generic forms always take place on the original source file.

CHECK <file>

The CHECK script file performs static analysis of a program unit using ISTLX, ISTYP, ISTSA and ISTPF. No provision is made for library input to ISTPF.

It should also be noted that the monolithic forms of the tools may also be used from ISTCE. This would allow, for instance, source-to-source polishing to be specified in the single command line:

```
LP SRC, POL, PL.OPT
```

2.4.   ISTCI – Object-oriented Interface

The Toolpack/1 experimental command interpreter provides a novel user interface that does not obey the normal 'verb'-'object' syntax. The command interface is object oriented in the sense that the actions to be performed are implicit in the name of the specified object. The object in question may be atomic, such as a file containing a Fortran unit or test data for example. It may also be a reference to set of objects, such as a suite of documentation files. Derived objects are regarded as views of an original object; for example a formatted version of a Fortran program is a view of the original program. All commands to ISTCI are requests to display a view of an object. If the specified view already exists as the result of a prior operation then it is displayed; if it does not exist then ISTCI will create it and display it. In effect the user may operate as if all possible views of all available objects exist at any time. The process of creating the requested view is controlled by a tool-object derived graph which, along with the state of the filestore of objects, is interrogated by the command interpreter, in determining what actions to take in order to fulfil the user's requested action. Thus the user is relieved of the sometimes considerable burden of marshalling all the necessary resources required to carry out an action such as conducting a semantic analysis of selected members of a suite of program units, for example the entire process is transparent to the user except for the computing time required.

The following are examples of view specifications for ISTCI:

a.f:pol	(the formatted view of the source file 'a.f')
a.ref:pol	(the formatted view of all the source files named in the reference file 'a.ref')
a.f:pol:scn	(the listing produced from lexical analysis of a polished version of 'a.f')
a.f:viw	(a symbol table listing derived from 'a.f' by ISTVS)

Further information on the Odin interpreter from which ISTCI was originally derived can be found in [7]. The operation may be likened to an interactive and enhanced form of the Unix MAKE facility.

The Toolpack/1 Tool Suite

Ian C. Hounam,
NAG Ltd
Oxford, U.K.

"Give us the tools, and we will finish the job."

— Winston L. S. Churchill,
Radio Broadcast, 9th February 1941

Abstract. In this section each Release 2 tool is at least summarised with a reference to the appropriate chapter of this book. Those not dealt with in other chapters are described in more detail.

1.   Command Executors

ISTCE   The Toolpack/1 Command Executor. Provides a closed environment to tool users and access to the Portable File Store. It can only run in the embedded TIE regime. (see Chapter "Toolpack Invocation Techniques")

ISTCI   An experimental object-oriented command interpreter. This is included as a demonstration of an Odin type of command interpreter. It is not intended as a working interface to the tools as the overhead on machine resources is too great. (see Chapter "Toolpack Invocation Techniques")

2.   General Tools

ISTDC   Data comparison tool for comparing files of numeric values, also handles embedded text. (see Chapter "Documentation and Non-Fortran Tools")

A. A. Pollicini (ed.), Using Toolpack Software Tools, 53–61.
© 1989 ECSC, EEC, EAEC, Brussels and Luxembourg.

ISTET   This tool removes the tab characters from a file. If the text option is selected the tab characters are replaced by a sufficient number of spaces to move to the next tab stop; tab stops are every 8 spaces (i.e. columns 9, 17, ...). If the input format is D (for Digital Equipment Corporation), the input is considered to be in DEC's "tab format" Fortran, and is converted to fixed-format Fortran.

ISTFI   This tool prints out all the include file names that a file needs, even if they are nested. The user can specify whether the PFS and/or the HFS are to be searched. The tool recognises TIEMAC or ISTMP include statements

include <file_name>

and the ISTRF switch input command

.so <file_name>

An option is available to allow the user to select TIE conforming names, that is search the PFS for files that do not start with the host file identifier #. Alternatively TIEMAC type host file names can be selected.

ISTGP   Searches multiple files for occurrences of a regular expression. A regular expression is a search string that can contain wildcards. Examples of a regular expressions are:

TIE?*   – which will find the words TIECODE, TIEMAC and any string starting with the letters TIE.

Run[0-9] – will find all occurrences of Run followed immediately by a digit.

Full details of regular expressions can be found in Section 3.2 of the Chapter "The Toolpack/1 Editor and Other Fortran Modifiers". This tool will print out all lines containing a regular expression, or alternatively all lines not containing the expression. Another option can select the reporting of only the names of files that have at least one line containing the expression.

ISTHP   This tool provides limited help information about tools. The user enters regular expressions which are matched against keyword lists and appropriate summaries of tool operation are listed. The information in the help system is based on that provided in Appendix B of *"Toolpack/1 Release 2 Introductory Guide"* [17].

Example HP-1

The entry for ISTDS would be found by specifying the following key words : **ISTDS, Declaration, Standardiser or DECS.**

The help output is as follows:

ISTDS - Declaration Standardiser

This tool standardises the declarative parts of Fortran 77
program units.

>   Pre-requisites:
>   ISTLX, ISTYP
>   Arguments:
>   <input parse tree (*)
>   <input symbol table (*)
>   <input comment index (*)
>   <input comment stream (*)
>   >output token stream (*)
>   >output comment stream (*)
>   [<Options] (*)

ISTSP   The TIE conforming version of the SPLIT utility, which breaks up a file using
         split markers that specify the individual files and their names. This tool
         recognises the same marker as the SPLIT program:

C$$$SPLIT$$$ <file_name>

It also can use a source embedded directive (SED) form of marker:

*SP <file_name>

The filenames on these split markers, because ISTSP is TIE conforming, are
assumed to be in the PFS, unless the host file identifier character #, is used.

ISTSV   This is a save/restore utility that can save and restore sub-trees of the PFS.
         This tool will concatenate files, for example a directory and sub-directories
         of the PFS, marking each file with an SED of the form:

*sv <file_name>

The file name contains the full directory path, hence the saved files can
subsequently be restored to the same directory structure. This tool is
useful for making backups of the PFS. The ISTCE tool file contains certain
Toolpack/1 examples and other useful files in an ISTSV file. These should
be restored to the PFS.

ISTTD   The text file comparison tool. (see Chapter "Documentation and Non-Fortran
         Tools")

ISTVC   This is a simple text file version controller. Versions may be set up and recalled
         either by version number or date/time. (see Chapter "Documentation and
         Non-Fortran Tools")

3.   Documentation Generation Tools

ISTAL   The documentation generation aid is a preprocessor that can be used to generate callgraphs, cross reference listings, segment execution frequencies and symbol information from intermediate files created by other tools. (see Chapter "Documentation and Non-Fortran Tools")

ISTCB   This is a comparison tool. Change bar requests are inserted into a document wherever it is found to differ from the original. (see Chapter "Documentation and Non-Fortran Tools")

ISTDX   This tool extracts marked documentation from a Fortran program. (see Chapter "Documentation and Non-Fortran Tools")

ISTRF   The Toolpack/1 text formatter. (see Chapter "Documentation and Non-Fortran Tools")

4.   Fortran 77 Oriented Tools

This class of tools is intended to manipulate Fortran 77 programs. However, in consideration of the variety of extensions usually implemented by Fortran 77 processors, some constraints are put on the language "understood" by these tools, which is referenced as **Toolpack/1 Target Fortran.**

Toolpack/1 tools accept as input ANSI standard Fortran 77 [1] plus some extensions. These extensions are additions to the language definition that can be found in many compilers. The portability verifier ISTPF, however, will report any use of these extensions. The main additions are listed below and full details may be found in [13].

(1) Long symbolic names are allowed and may contain *underscore* and *dollar sign.*

(2) Comments may start with any non-numeric character.

(3) Additional data types are allowed: DOUBLE COMPLEX, INTEGER*2, LOGICAL*1, etc.

(4) REAL*8 is recognised as DOUBLE PRECISION.

(5) Additional intrinsic functions to support additional types are provided.

ISTAN   This is an instrumentor for Fortran 77 that allows the collection of run time execution frequency information. (see Chapter "Analysis Tools for Fortran 77")

ISTCN   Fortran name changer (token stream level). (see Chapter "The Editor and Other Fortran Modifiers")

ISTCR   Fortran name changer (symbol table level). (see Chapter "The Editor and
        Other Fortran Modifiers")

ISTDS   The Toolpack/1 declaration standardiser.   (see Chapter "Fortran 77
        Transformers")

ISTED   The Fortran aware editor. (see Chapter "The Editor and Other Fortran
        Modifiers")

ISTFD   This is a Fortran program comparison tool that works on token streams
        rather than source text. ISTFD will report as identical, programs that are
        functionally the same, but contain formatting differences and changes in
        the comments. For example ISTFD can compare a program that has had
        format changes made using the polisher with the original, reporting only
        changes significant to the running of the program. This tool is considerably
        more useful to the Fortran programmer than the standard text differencing
        programs found on most systems.

Example FD-1

The following two programs would be found identical by ISTFD:

```
      C This program ...
            PROGRAM TEST
            DO 10 I=1,10
      C Output the index
            PRINT *,I
        10 CONTINUE
            STOP
            END
```

```
            PROGRAM TEST
            DO 10 I= 1, 10
            PRINT *,I
      C A comment
        10 CONTINUE
      C End of program
            STOP
            END
```

ISTFP   A fast unscanner. ISTFP performs a conversion of a token and comment
        stream to Fortran 77. This is carried out very efficiently, but to a fixed
        format, and with the loss of comments. Can be used to output a program,
        that has been processed by Toolpack/1 tools, for compilation. The output
        program would not normally be kept. It might be used after, for example,
        the DO loop unroller tools had been used.

ISTFR   Converts the format of real, double precision and complex constants to a
        standard form. ISTFR will:

        (1) Convert exponent character to upper case.

        (2) Remove any '+' on the exponent.

        (3) Remove a zero exponent.

(4) Change, if necessary, the mantissa to contain one digit before and after the decimal point.

Example FR-1

Input   Output

```
.1234E+0                                0.1234
.567e+10                                5.67E9
36.12E6                                 3.612E7
```

ISTGI   This tool will change all Fortran 77 intrinsic function references to their generic forms (where this is legal or possible).

ISTIN   Inserts the contents of include files into a file. Include files are specified by a statement of the form

`include <file_name>`

The bounds are marked with source embedded directives of the form:

`*$in$ begin <file_name>`
`*$in$ end`

This tool is similar to the include file function of TIEMAC, but the included files are delimited by markers, so that the tool ISTUN can reverse its operation. The declaration standardiser ISTDS, can optionally be set not to process such included files. An option can be set to specify that the HFS should be searched for include files, irrespective of the presence of the host file identifier #, rather than the normal TIE convention.

ISTJS   Joins strings in FORMAT statements, can optionally convert holleriths to strings and can convert X descriptors.

ISTLS   This tool changes long Fortran names to the ANSI standard of 6 characters.

ISTLX   The Toolpack/1 lexer (or scanner). (see Chapter "Analysis Tools for Fortran 77")

ISTME   Manipulates Fortran expressions to reduce the depth of the intermediate value stack.

ISTMF   This is a Fortran differencing program. It takes as input two programs, one in source form, the other in token stream form and produces a merged output source. Inserted lines are marked with comments at the beginning and end of each. Deletions and changes are included in the output file as comment lines, suitably marked.

Example MF-1

```
                PROGRAM TEST

                READ *,I
    * $mf$ Changed From
    * IF (I.NE.2)GO TO 10
    * $mf$ Changed To
                IF (I.NE.1) GO TO 10
    * $mf$ End Change
                Y = 10.0
                Z = X*Y
                GO TO 20

      10 X = 1.0
    * $mf$ Deletion
    * X=X*X
    * $mf$ End Deletion
                Z = X/Y
      20 CONTINUE
    * $mf$ Insertion
                PRINT *,Z
    * $mf$ End Insertion
                END
```

ISTMP   Provides a similar function to TIEMAC, but is TIE conforming.

ISTPF   The Fortran 77 portability verifier (based on PFORT). (see Chapter "Analysis Tools for Fortran 77")

ISTPL   The Fortran 77 polisher. (see Chapter "The Fortran 77 Source Polisher")

ISTPO   Polish options editor, a menu driven editor that will create or modify a file of ISTPL options. (see Chapter "The Fortran 77 Source Polisher")

ISTPP   Ensures the consistency of PARAMETER statements over Fortran program units

ISTPT   The Toolpack/1 Fortran precision converter. (see Chapter "Fortran 77 Transformers")

ISTSA   Fortran 77 semantic analyser. (see Chapter "Analysis Tools for Fortran 77")

ISTST   Fortran 77 structurer. (see Chapter "Analysis Tools for Fortran 77")

ISTUN   Reverses the effect of ISTIN. Removes included files which are marked by SEDs, placing include statements instead. The SED, previously inserted by ISTIN, contains the file name referred to by the generated include statement.

ISTYP   The Toolpack/1 parser. (see Chapter "Analysis Tools for Fortran 77")

ISTYF   This is a parse tree flattener. It reverses the function of ISTYP, recreating the token and comment stream. It is used after a tool that works from symbol table to symbol table, e.g. ISTGI. ISTYF produces a token/comment stream that can be subsequently polished to give Fortran 77 output.

5.   Monoliths

Eight combined tool sequences arc available to carry out common applications of the tools. These monolithic tools are much faster than the equivalent sequence of tool fragments, and they eliminate the need for intermediate files. The following table shows the monoliths and the equivalent tool fragment sequences.

Monolith	Tool sequence	Function
ISTLA	ISTLX/ISTYP/ISTSA	Semantic Analysis
ISTLP	ISTLX/ISTPL	Polish
ISTLY	ISTLX/ISTYP	Parse
ISTP2	ISTLX/ISTPP/ISTPL	Standardise Parameters
ISTQD	ISTLX/ISTYP/ISTDS/ISTPL	Declaration Standardisation
ISTQP	ISTLX/ISTYP/ISTPT/ISTPL	Precision Transformation
ISTQT	ISTLX/ISTYP/ISTPT	
ISTDT	ISTYP/ISTDS/ISTPL	

The first six monoliths in the table work source-to-source, while the other two do not. However, the final two are designed so that, when used in sequence, precision transformation is combined with declaration standardisation and the whole process is performed source-to-source.

6.   DO Loop Unrolling Tools

These three tools can restructure DO loops in a form that executes faster on vector processors. (see Chapter "DO Loop Transforming Tools")

ISTCD   This tool combines consecutive DO loops.

ISTSB   Recombines and substitutes expressions.

ISTUD   Unrolls DO loops.

7.   Experimental Tools

ISTJF   A source level differentiator that will create a derivative function from the
        analysed source. This tool recognises and produces Fortran 66 code. Input
        to ISTJF is a single subroutine that contains a CONSTRUCT command to
        ISTJF that defines the derivation to be produced. Examples of this command
        are:

        CONSTRUCT D(Y)/D(X) IN GRAD(N)
        CONSTRUCT D(F)/D(X) IN JACOB(M,N)

        The first of these examples instructs ISTJF to construct a routine that will
        calculate the gradient of the function in the array. The second example
        calculates the Jacobian. The results can either be in single or double
        precision, depending on the run time library selected.

ISTX1

ISTX2

ISTX3   These are three small expert systems written in Fortran. These are based on
        programs by Chris Naylor [43]. They are included as a demonstration of TIE
        conforming programs.

Part II
Use of Selected Tools

"With a point to stand upon, "he" could move the world."

— attributed to Archimedes, third century BC

"Na grama lavandera treuva mai la bon-a pera"

— Piedmont proverb arguing the counter thesis that *no tool is good in unskilled hands*

Analysis Tools for Fortran 77

Malcolm J. Cohen
NAG Ltd
Oxford, U.K.

> *"You see it's like a portmanteau – there are two meanings packed up into one word."*
>
> — Lewis Carroll,
> Through the Looking-Glass, ch. VI, 1872

Abstract. This chapter provides an overview of the major Fortran 77 analysis tools included in the second release of Toolpack/1. These tools provide the Toolpack/1 user with comprehensive static checking and conformance checking capabilities.

1.   Introduction

This chapter discusses the design, implementation and use of the four major Fortran 77 analysis tools included in Toolpack/1 Release 2. These tools provide a sequence of analyses of increasing sophistication (and cost) which may be used to provide the tool user with the desired level of analysis.

The stages of analysis provided are lexical analysis, syntax analysis, static semantic analysis and portability verification. The output from the first two analysis stages are essentially alternate representations of the user's program, being a token stream and abstract syntax tree respectively. The third stage annotates the output from the second stage with globally derived static information, whilst the only output from the fourth stage is an error report.

2.   Lexical Analysis

Lexical analysis of a piece of text is the process of recognising its textual (or lexical) elements. These elements are the words and punctuation from which it is consituted,

A. A. Pollicini (ed.), Using Toolpack Software Tools, 65–99.
© 1989 ECSC, EEC, EAEC, Brussels and Luxembourg.

and are called *tokens*. Each token has a classification (e.g. word, number, left parenthesis, ...) or "token type" and possibly a value (i.e. its textual representation) – some tokens may not need a value as the token type may completely specify it (e.g. left parenthesis).

Thus the input to a lexical analyser may be thought of as a stream of characters (including end-of-line characters) and its output as a stream of tokens, or token stream.

2.1.   Fortran 77 Lexical Characteristics

The characteristics of Fortran 77 are such that lexical analysis is harder than for most other languages. The two most troublesome characteristics are insignificant blanks and non-reserved keywords.

Insignificant blanks prevent the delimiting of keywords and user-defined names, for instance, by blanks. This means that to tell the difference between 'DO100I=' being two tokens (<name> + <equals sign>) or being four tokens (<DO keyword> + <label> + <name> + <equals sign>) the lexical analyser must look much further along the line; in fact there may be an arbitrary amount of text to be examined (within the limit of 19 continuation lines) before the decision can be made, and while performing this look-ahead the scanner must "balance" quotes and parentheses.

Non-reserved keywords add to this difficulty. Again the scanner must look ahead an arbitrary amount of text. An example is the logical IF statement; the scanner cannot choose between "IF" being a keyword or a name until it has found the closing parenthesis and seen whether it is followed by an equals sign or not.

2.2.   The tool ISTLX

The Toolpack/1 scanner, ISTLX, takes as input a single file containing Fortran 77 source text. The output consists of error and warning messages, an optional list file and at least one token stream. Each token stream is made up of a single token file and a single comment file. The scanner may produce multiple token streams from a single input file, each output token stream containing a single program-unit: see the ISTLX documentation [36] for further details.

*   **The Token Stream**

The token stream produced by ISTLX is described in detail in the first Appendix to this chapter. Here we will simply note the three special tokens: TCMMNT, TZEOS and TZEOF. TCMMNT is the comment token, and may appear anywhere in the token stream. TZEOS is the end-of-statement token, and signals the end of a Fortran statement. This token is emitted by ISTLX whenever the next non-comment line in the input file is the initial line of a Fortran statement (i.e. it is not a continuation line). TZEOF is the end-of-file token, and is always the last token in the stream.

- **Internal Operation**

ISTLX is a table-driven scanner created by the FSCAN [6] system. It overcomes the difficulties of scanning Fortran (mentioned in Section 2.1) by making guesses as to what type of statement is being read and buffering the output tokens until the guess is confirmed or denied; if the guess was wrong ISTLX discards the buffered output and rescans from the point where the guess was made.

The advantages of using a table-driven scanner instead of a hand-written one are:

(1) the scanner may easily be changed if there is a change to the input language,

(2) the input to the table generator should be easier to understand than the equivalent Fortran code, and so easier to maintain.

The main disadvantage of this approach is that it is somewhat slower than a good hand-written scanner.

Example LX-1

Input statement:
```
        X = 3.3+5/4
```

Output token stream:
```
        TNAME    "X"
        TEQUAL
        TRCNST   "3.3"
        TPLUS
        TDCNST   "5"
        TSLASH
        TDCNST   "4"
        TZEOS
```

3. Syntax Analysis and Transformational Grammars

Syntax is the textual or symbolic structure of a language. The (usually finite) specification of this structure is called a grammar; a grammar defines the legal sentences of a language without specifying what they mean.

A transformational grammar is a (finite) formal specification of a potentially infinite language. A language is considered to be the set of all strings which are legal "sentences" in the language. The grammar consists of:

(1) the set of terminal symbols (vocabulary),

(2) a set of non-terminal symbols, one of which is the "start symbol",

(3) a set of rewriting rules, called productions. These have the general form

$$a \rightarrow b$$

where a and b are strings of terminals and non-terminals.

The language specified by such a grammar is the set of all strings containing only terminal symbols which may be derived from the start symbol using the productions.

3.1.   Classes of Grammar

According to limitations placed on the productions, different classes of languages may be described. There are four such classes, each more restrictive than the previous class. The most general class is type 0 or "unrestricted", and results when there are no restrictions on the form of the productions.

The next most general class is type 1 or "context-sensitive"; in this class the productions are constrained to be of the form

$$x \, A \, y \rightarrow x \, z \, y$$

where x, y and z may be arbitrary strings of terminals and non-terminals, but A must be a single non-terminal symbol.

The third class (type 2) is called "context-free". The productions are constrained to be of the form

$$A \rightarrow x$$

where A is a single non-terminal and x is a string of terminals and non-terminals.

The most restrictive class is type 3, or "regular". A regular grammar must be either "right linear" or "left linear". In a right linear grammar the productions must be of the form

$$A \rightarrow x$$
$$\text{or} \quad A \rightarrow x \, B$$

where A and B are non-terminal symbols and x is a string of terminal symbols. In a left linear grammar the productions must be of the form

$$A \rightarrow x$$
$$\text{or} \quad A \rightarrow B \, x$$

3.2.   Parsing a Context-free Grammar

Parsing is the process of discovering the productions which must be applied to the start symbol in order to derive the target sentence. The derivation may be treated as a tree with the start symbol as the root, since each production rewrites exactly one non-terminal. This derivation tree is complete when all its leaves consist of the terminal symbols which then make up the sentence. This derivation tree is also known as the syntax tree or parse tree (when it is created by parsing).

The class of context-free grammars may be further subdivided according to how difficult it is to produce the correct derivation tree from an input sentence, assuming the parse procedes from left to right. A very general class is LR(1), in which the derivation detected is determined by the left context (i.e. what we have recognised so far), the phrase being reduced plus one symbol (token) of right context (look-ahead). An LR(1) grammar may be parsed deterministically by the LR(1) algorithm; that is, no backtracking is ever needed.

ISTYP uses an LALR(1) algorithm, which is almost as powerful as LR(1). Like LR(1), LALR(1) has the properties of linearity (the time taken is proportional to the length of the input) and immediate error detection (an error is detected at the first symbol which cannot be a valid continuation of the input).

3.3.   The tool ISTYP

The Toolpack/1 parser, ISTYP, takes as input a single token stream (consisting of a token file and comment file). Its output consists of error and warning messages, a parse tree, a string table, a symbol table and a comment index.

*   **The Input Language**

ISTYP accepts standard conforming Fortran 77 [1] with the Hollerith extension as described in Appendix C of [1] and additional datatypes and intrinsic functions described in *"Toolpack/1 Target Fortran 77"* [13]. It is a table-driven parser with the tables generated using the YACC [40] parser-generator. The second Appendix to this chapter contains a listing of the pure grammar, in YACC format, which describes the source language.

The grammar used by ISTYP is actually slightly less restrictive than the above description implies. Some of this is due to the difficulty in parsing Fortran with only one symbol of look-ahead:

(1) A complex constant will accept an implied DO-loop or expression as its first part. This is because both of the following are legal Fortran:

```
PRINT *,(5,3,I=1,10)
PRINT *,(5,3)
```

The parser does not know whether it has a sequence of expressions or a complex constant until the second comma or final parenthesis has been read. This can theoretically be avoided but only at the cost of increasing the size (and decreasing the speed) of the parser.

(2) The file specification for BACKSPACE, ENDFILE and REWIND may include an implied DO loop as the unit-identifier. This is partly because the parser doesn't

have a separate set of productions for recognising integer expressions, and because of the previous problem. Again, this could be avoided by increasing the size of the parser.

One area of difficulty is due to a syntactic ambiguity in Fortran. This is the statement:

READ (NAME)

If NAME is a character item, then the statement is a formatted READ of the form "READ <format-id>" (from the default input device); if NAME is of type integer however, it is an unformatted READ of the form "READ (<unit-id>)" from the unit number thus specified.

Finally, there are a number of things which are not checked by the parser because:

(a) they are not particularly syntactic in nature,

(b) they are common extensions which have no or little impact for tools which use the parser output, or

(c) due to the cost of checking in size, speed and complexity of the parser it is more appropriate that they appear in the static semantic analyser.

Some of these are:

(1) Arrays may have any number of dimensions.

(2) Dummy arguments may appear in EQUIVALENCE statements.

(3) Statements may appear in any order except for PROGRAM, FUNCTION, SUBROUTINE, BLOCK DATA and END which must appear initially or terminally as appropriate.

(4) The value of N (which is an integer constant) in "INTEGER*N" (and other such forms) may be any digit string.

All of these "gaps" in ISTYP are filled in by the static semantic analyser, ISTSA.

- **The String Table**

As each token with a value (text) is read by the front end of the parser, its text is placed into the string table. The value of the token then passed to the main body of the parser is the appropriate index into the string table. This reduces the amount of storage required, since each string is only stored once regardless of how many times it occurs throughout the program file, and eliminates the need to pass strings around inside the parser. A hash table index is used during construction of the string table for efficiency.

- **The Symbol Table**

The symbol table contains a list of all discrete symbols encountered during the parse of the program file together with some attribute information. There is a program-unit symbol for each program-unit in the file; this need not be the first symbol of the program-unit however. Most symbols have pointers to them in the parse tree, the only exceptions being an unnamed main program (the symbol has the name "$MAIN", an unnamed BLOCKDATA subprogram (the symbol has the name "$BLOCKDATA" and an unnamed common block (i.e. blank common; the symbol has the name "$COMMON").

The detailed structure of the symbol table is described in " *ISTYP – Toolpack/1 Parser*" [9].

- **The Parse Tree**

The "parse tree" produced by ISTYP is not actually a simple parse tree but an abstract syntax tree. An abstract syntax tree differs from a parse tree in that most superfluous structure has been discarded, leaving a more compact and convenient object. The detailed structure of the parse tree is described in *"ISTYP – Toolpack/1 Parser"*. An example of the parse tree of the statement :

$$X = 3.3+5/4$$

is given in Figure 1.

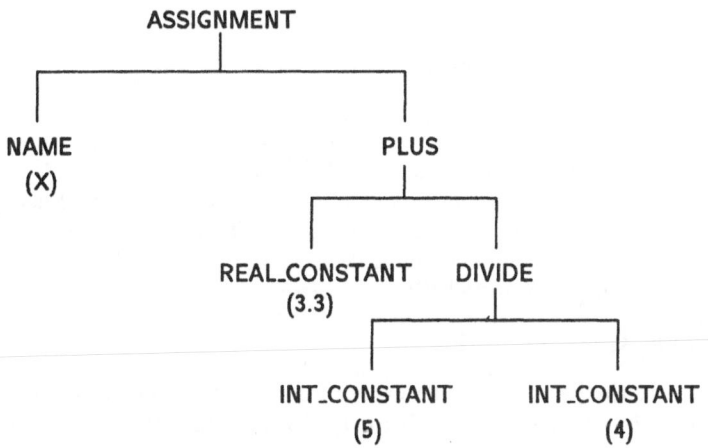

Figure 1. Diagram YP-1

- **The Comment Index**

The comment index created by ISTYP allows the comments associated with each statement to be retrieved from the comment file without needing to read the token stream. A comment is "associated" with a statement if it occurs immediately before the initial line or a continuation line of the statement.

The comment index thus records the statement number of each statement which has comments together with the first and last comment line numbers in the comment group.

Statements are numbered beginning with 1 as the first statement of the first program-unit in the parse tree. Comment lines are similarly numbered with 1 beginning the first comment line in the program file.

4. Static Semantic Analysis

Static semantic analysis as performed in Toolpack/1 consists in the following tasks:

(1) The local symbol table has more information added,

(2) The parse tree nodes of expressions are annotated,

(3) Semantic checking is performed,

(4) The global symbol table is created.

Static semantic analysis should find all language errors in a program which do not depend on dynamic semantics (e.g. flow of control).

4.1.   The Tool ISTSA

The Toolpack/1 static semantic analyser, ISTSA, takes as input the parse tree and symbol table produced by the parser (ISTYP). Its output consists of error and warning messages and an attributed parse tree: the latter consists of modified parse tree and symbol table files together with an attribute file. These modified files may safely replace the original parse tree and symbol table files, as the modifications do not affect tools which use the output from ISTYP.

●   **Parse Tree Annotation**

Each node in the parse tree of an expression defines a subexpression, consisting of all the nodes below that level (see Figure 1). ISTSA annotates each such node with the following information:

(1) datatype of the subexpression,

(2) whether the subexpression is an array,

(3) whether the subexpression is a procedure,

(4) whether the subexpression is constant,

(5) value of an integer constant subexpression,

(6) length of a character subexpression (zero if unknown).

See Figure 2 for an example of annotation. This annotation is used in the determination of array sizes, equivalences, etc. where integer constant expressions may be used.

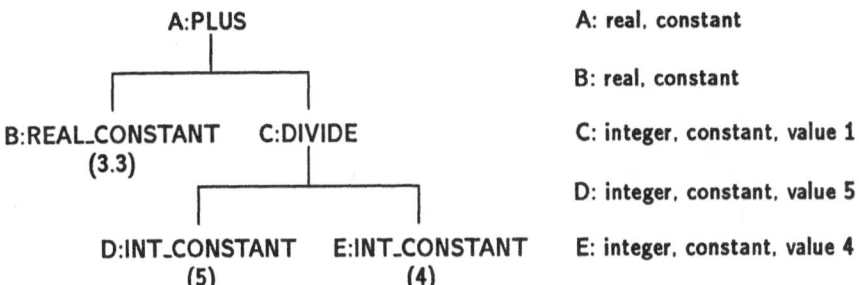

Figure 2. Diagram SA-1

- **Some Examples of Static Semantic Analysis**

Following are two examples of ISTSA usage. The first example is of a program with some simple semantic errors; ISTSA produces an error report (and no output files). The second example is of a semantically correct program, and a description of the extended symbol table data created by ISTSA.

Example SA-1

Input:

```
SUBROUTINE EX1(SA)
INTEGER I(10),J(10)
CHARACTER SA
EQUIVALENCE (SA,I)
EQUIVALENCE (A,B,C,I(5)),(J(9),A)
EQUIVALENCE (I(3),J)
COMMON A
END
```

Output:

```
Error: Dummy argument in EQUIVALENCE at statement 4 in EX1
Error: Inconsistent EQUIVALENCEs at statement 6 in EX1
[EX1 processed]
[No global processing]
[ISTSA Terminated, Errors detected]
```

Example SA-2

Input:

```
PROGRAM EX2SA
COMMON/FRED/I(10)
I(3)=5
CALL ASUB(3)
END
SUBROUTINE ASUB(K)
COMMON/FRED/D
DIMENSION J(6),I(5)
EQUIVALENCE(I(5),J),(I,D)
PRINT *,I(K)
END
```

Symbol Table Data Created by ISTSA:

Local Symbol Table:

```
Program Unit: EX2SA PROGRAM
        Common blocks:
            FRED, Size: 40
                Items: I
                Usage: Assigned to on left of "="
        Variables:
            I INTEGER, Array (10),
              in common block /FRED/ with offset 0

Program Unit: ASUB SUBROUTINE
        Common blocks:
            FRED, Size: 40
                Items: D
                Usage: In an expression
        Variables:
            D REAL, in common block /FRED/, offset 0
            I INTEGER Array (5),
            equivalenced into common block /FRED/, offset 0
            J INTEGER Array (6),
            equivalenced into common block /FRED/, offset 16
```

Global Symbol Table:

    Program Units

        EX2SA: PROGRAM
            Common Block /FRED/, updated
            Calls ASUB

        ASUB: SUBROUTINE
            Argument 1: INTEGER , read-only
                Actual arguments: expression (from EX2SA, statement 4)
            Common Block /FRED/

    Common Blocks

        /FRED/, Length 40, non-character

5. Approach to Portability Verification

The portability verification provided by the ISTPF tool consists of three facets:

(1) Checking for conformance to the Fortran 77 standard, in particular the areas in which compilers are generally unhelpful. The major part of this is in the communication between program-units.

(2) Checking for avoidance of the less portable parts of Fortran 77. To this end, a more portable subset (called PFORT-77) is defined.

(3) Checking of unsafe references and common usage.

5.1. The tool ISTPF

The Toolpack/1 portability verifier, ISTPF, takes as input at least one attributed parse tree produced by ISTSA, together with any number of "library" attribute files (also produced by ISTSA). The "library" attribute files are used only for inter-program-unit checks; only the program-units for which the entire annotated parse tree is available will have local portability verification checks applied. The only output from this tool is a set of error and warning messages.

5.2. Fortran 77 Conformance Checking

ISTPF checks for the following deviations from the Fortran 77 standard in the user's program and reports an error if found.

(1) Incorrect or inconsistent number of arguments to an external subprogram.

(2) External subprogram referenced with the wrong (or an inconsistent) datatype. This includes a function being called as a subroutine.

(3) Incorrect or inconsistent datatypes or structures (e.g. array, variable or procedure) of arguments to subprograms.

(4) Inconsistent common block sizes.

(5) Common block containing both numeric and character items.

(6) Inconsistent "SAVE"-ing of common blocks. (Any common block must either *always* or *never* be SAVEd.)

(7) A format-specifier must not be an INTEGER, REAL or LOGICAL array (these are part of the Hollerith extension).

(8) Variables referenced but never set (i.e. those which appear in an expression but do not appear on the left-hand side of an assignment statement, nor are they used in DATA, ASSIGN, or as the actual argument in a CALL or function reference).

(9) Function value never set.

Additionally, ISTPF checks for instances of "clutter" (unused items) and produces a warning message if found. This includes:

(1) Unused dummy arguments, local variables and common blocks,

(2) Local variables which are set but never referenced.

5.3.   The PFORT-77 Subset Language

The following describes the difference between standard Fortran 77 and the portable subset accepted by ISTPF.

(1) A COMMON block must not appear in more than one BLOCK DATA subprogram

(2) Within a COMMON block, any COMPLEX and DOUBLE PRECISION entities must precede all other types.

(3) EQUIVALENCE between items of differing type must not equivalence non-initial elements of arrays.

(4) Intrinsic functions used as actual arguments must be declared INTRINSIC. (This is actually checked in ISTYP rather than ISTPF).

(5) Non-standard intrinsic functions must not be used.

(6) Intrinsic functions must not be explicitly typed.

(7) External functions must be explicitly typed.

(8) A data-value must be exactly the same type as the corresponding data-item.

(9) A variable used in an array-declarator must not have its type declared following the occurrence of the array-declarator.

(10) A DO loop must have an INTEGER control variable.

(11) There is no PAUSE statement.

(12) An assigned GOTO must not have a label list.

(13) The T and TL edit descriptors must not be used.

(14) The only relational operators to be used on character strings are .EQ. and .NE. (the intrinsic functions LLE, LLT, LGE, LGT being used instead of the other relational operators).

(15) Within each program-unit, the names of common blocks, statement functions, dummy arguments, statement function dummy arguments and variables which appear in ASSIGN statements must all be distinct.

(16) The name of a function subprogram must not be used as the control variable of a DO loop within that subprogram.

(17) The control variable of a DO loop must not appear in the limit expressions for that loop.

(18) The maximum length of a CHARACTER variable is 255.

(19) The maximum length of a character constant is 64.

(20) Non-standard control-list items (in I/O statements) must not be used. (This is actually checked by ISTSA, which produces a warning message).

Additionally, deviation from the following conditions results in a warning message:

(1) All external references must be declared EXTERNAL.

(2) A variable used in an array-declarator must not be modified by that routine.

(3) Only the standard character set, excluding the currency symbol, may be used in character constants.

5.4.　　Unsafe References

The Fortran 77 standard does not specify the exact mechanism to be used for argument passing in a program, instead prohibiting the program from doing anything which might reveal what the mechanism is. The two most commonly implemented mechanisms are "call by reference" and "call by value-result". In call by reference, each reference to the dummy argument references the actual argument. In call by value-result, the dummy argument is treated as a local variable which is initialised to the value of the actual argument; the actual argument receiving the result (final value of the dummy argument) upon termination of the subprogram.

In order to prevent the user program from uncovering these mechanisms, the standard specifies that under certain conditions, the dummy argument must not be "defined" (i.e. receive a value). ISTPF calls these "unsafe references" since the programs only deviate from the standard if the "definition" actually occurs at execution time; this is not in general determinable by static checking.

There are five types of unsafe reference diagnosed by ISTPF:

(1) Constant or expression associated with a dummy argument which may be changed.

(2) Actual argument associated with two dummy arguments, at least one of which is scalar, and either or both of which are changed.

(3) The actual argument is in a common block which is accessed either directly or indirectly by the called subprogram and either the dummy argument or the common block is modified.

(4) The actual argument is the dummy argument of a statement function and the associated dummy argument (in the called function) may be changed.

(5) The actual argument is an active DO-loop index and the associated dummy argument may be changed.

ISTPF treats an array as a single item. This may lead to an unsafe reference being signalled unnecessarily. To attempt to differentiate between apparently unsafe array references and other types, ISTPF will give the message a severity of "unsafe" rather than "error" if one of the receiving dummy arguments is an array.

5.5.   Common Block Usage

COMMON blocks are checked to ensure that either:

(1) The COMMON block is SAVEd, or:

(2) there is an "owning" program unit containing the common block which is always active whenever any other program-unit containing the common block is active.

This prevents the "undefinition" of a common block between references to two different subprograms which are using it for communication.

6.   Conclusions

We have described a sequence (shown in Figure 3) of analysis tools which provide the Toolpack/1 user with sophisticated Fortran 77 analysis facilities. These tools may be used either as a checking facility or for the provision of information to other tools. Other tools which are supported by this analysis suite range from documentation aids to program transformers.

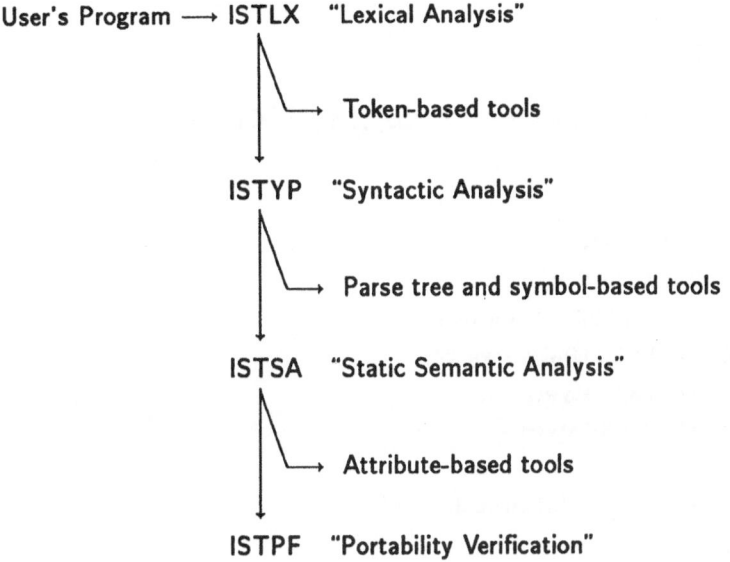

Figure 3. The Analysis Suite

Appendix:
Fortran 77 Token Types (ISTLX)

TZEOF	End of source file.
TASSIG	The ASSIGN keyword.
TBACKS	The BACKSPACE keyword.
TBLOCK	The BLOCKDATA keyword.
TCALL	The CALL keyword.
TCLOSE	The CLOSE keyword.
TCOMMO	The COMMON keyword.
TCONTI	The CONTINUE keyword.
TDATA	The DATA keyword.
TDO	The DO keyword.
TDIMEN	The DIMENSION keyword.
TELSE	The ELSE keyword.
TELSIF	The ELSEIF keyword.
TEND	The END keyword.
TENDFI	The ENDFILE keyword.
TENDIF	The ENDIF keyword.
TENTRY	The ENTRY keyword.
TEQUIV	The EQUIVALENCE keyword.
TEXTER	The EXTERNAL keyword.
TFUNCT	The FUNCTION keyword.
TFORMA	The FORMAT keyword.
TGOTO	The GOTO keyword.
TIF	The IF keyword.
TIMPLI	The IMPLICIT keyword.
TINQUI	The INQUIRE keyword.
TINTRI	The INTRINSIC keyword.
TOPEN	The OPEN keyword.
TPARAM	The PARAMETER keyword.
TPAUSE	The PAUSE keyword.

TPRINT　The PRINT keyword.

TPROGR　The PROGRAM keyword.

TREAD　The READ keyword.

TRETUR　The RETURN keyword.

TREWIN　The REWIND keyword.

TSAVE　The SAVE keyword.

TSTOP　The STOP keyword.

TSUBRO　The SUBROUTINE keyword.

TTHEN　The THEN keyword.

TTO　The TO keyword.

TWRITE　The WRITE keyword.

TINTEG　The INTEGER keyword.

TREAL　The REAL keyword.

TDOUBL　The DOUBLEPRECISION keyword.

TCOMPL　The COMPLEX keyword.

TLOGIC　The LOGICAL keyword.

TCHARA　The CHARACTER keyword.

TDCMPL　The DOUBLECOMPLEX keyword.

TCOMMA　A comma.

TEQUAL　An equals sign.

TCOLON　A colon.

TLPARN　A left parenthesis.

TRPARN　A right parenthesis.

TLE　A .LE. operator.

TLT　A .LT. operator.

TEQ　A .EQ. operator.

TNE　A .NE. operator.

TGE　A .GE. operator.

TGT　A .GT. operator.

TAND　A .AND. operator.

TOR　A .OR. operator.

TEQV　A .EQV. operator.

TNEQV　A .NEQV. operator.

TNOT　A .NOT. operator.

TSTAR　An asterisk (not when part of an exponentiation operator).

TDSTAR　A double asterisk (exponentiation operator).

    TPLUS   A plus sign.

    TMINUS   A minus sign.

    TSLASH   A slash (not when part of a concatenation operator).

    TCNCAT   A concatenation operator (double slash).

    TDCNST   A digit string (label or unsigned integer).

    TLCNST   A logical constant (.TRUE. or .FALSE.).

    TRCNST   A real constant.

    TPCNST   A double-precision constant.

    TCCNST   A character constant (note that the value of this token is the character constan
             value, i.e. it does not include the opening and closing quotes, etc.).

    THCNST   An Hollerith constant (the value of this token does not include the prefixed
             character count and H).

    TNAME   A symbolic name (not a Fortran keyword or format descriptor, but could
             be a control-information keyword (except for FMT, ERR and END) in I/O
             statements).

    TFIELD   A format edit descriptor, including a prefixed repeat count but not a prefixed scale
             factor.

    TSCALE   A format scale factor (P edit descriptor).

    TZEOS   End of the current statement.

    TCMMNT   A comment line.

    TFMTKD   The FMT specifier in a control-info list (I/O statement).

    TENDKD   The END specifier in a control-info list.

    TERRKD   The ERR specifier in a control-info list.

Appendix:
Fortran 77 Grammar (ISTYP)

```
%token TZEOF  TASSIG TBACKS TBLOCK TCALL  TCLOSE TCOMMO TCONTI TDATA  TDO
%token TDIMEN TELSE  TELSIF TEND   TENDFI TENDIF TENTRY TEQUIV TEXTER TFUNCT
%token TFORMA TGOTO  TIF    TIMPLI TINQUI TINTRI TOPEN  TPARAM TPAUSE TPRINT
%token TPROGR TREAD  TRETUR TREWIN TSAVE  TSTOP  TSUBRO TTHEN  TTO    TWRITE
%token TINTEG TREAL  TDOUBL TCOMPL TLOGIC TCHARA TDCMPL TCOMMA TEQUAL TCOLON
%token TLPARN TRPARN TLE    TLT    TEQ    TNE    TGE    TGT    TAND   TOR
%token TEQV   TNEQV  TNOT   TSTAR  TDSTAR TPLUS  TMINUS TSLASH TCNCAT TDCNST
%token TLCNST TRCNST TPCNST TCCNST THCNST TNAME  TFIELD TSCALE TZEOS  TCMMNT
%token TFMTKD TENDKD TERRKD

/**
 * Naming convention: all uppercase = terminal symbols (tokens)
 *                    all lowercase = non-terminal symbols which occur
 *                                    in the Fortran 77 standard
 *                    mixed case    = other non-terminal symbols
 **/

/* Precedence relations used to resolve expressions
   (lowest precedence listed first) */
%left TEQV TNEQV
%left TOR
%left TAND
%left TNOT
/* TLT etc not listed as they are not produced ambiguously */
%left TCNCAT
%left TPLUS TMINUS
%left TSTAR TSLASH
%right TDSTAR

%start executable_program

%%

executable_program : Program_unit
                   | executable_program Program_unit
                   ;
```

```
Program_unit         : main_program
                     | function_subprogram
                     | subroutine_subprogram
                     | block_data_subprogram
                     ;

main_program         : Optional_label program_statement Program_unit_block
                     | Program_unit_block
                     ;

function_subprogram  : Optional_label function_statement Program_unit_block
                     ;

subroutine_subprogram : Optional_label subroutine_statement Program_unit_block
                     ;

block_data_subprogram : Optional_label block_data_statement Program_unit_block
                     ;

Optional_label  : /* empty */
                | Label
                ;

Program_unit_block  : Statements End_PU
                    | End_PU
                    ;

Statements  : Statement
            | Statements Statement
            ;

Statement   : Optional_label Statement_body TZEOS
            | Label format_statement TZEOS
            | Optional_label error TZEOS
            /* error-handling: recover to the end of the current statement */
            ;

End_PU     : Optional_label TEND TZEOS ;

Statement_body  : entry_statement
                | parameter_statement
                | implicit_statement
                | data_statement
                | dimension_statement
                | equivalence_statement
                | common_statement
                | type_statement
                | external_statement
                | intrinsic_statement
```

```
                    | save_statement
                    | do_statement
                    | logical_if_statement
                    | block_if_statement
                    | else_if_statement
                    | else_statement
                    | end_if_statement
                    | assignment_statement
                    | Assignment_or_stmt_fn
                    | goto_statement
                    | arithmetic_if_stmt
                    | continue_statement
                    | stop_statement
                    | pause_statement
                    | read_statement
                    | write_statement
                    | print_statement
                    | rewind_statement
                    | backspace_statement
                    | endfile_statement
                    | open_statement
                    | close_statement
                    | inquire_statement
                    | call_statement
                    | return_statement
                    ;

/*******************/

program_statement    : TPROGR Name TZEOS ;

/*******************/

function_statement : Data_type TFUNCT Name Fplist TZEOS
                    | TFUNCT Name Fplist TZEOS
                    ;

Fplist             : TLPARN Fp_namelist TRPARN
                    | TLPARN TRPARN
                    ;

Fp_namelist        : Name
                    | Fp_namelist TCOMMA Name
                    ;

/*******************/
```

```
subroutine_statement: TSUBRO Name TZEOS
                    | TSUBRO Name TLPARN TRPARN TZEOS
                    | TSUBRO Name TLPARN Splist TRPARN TZEOS
                    ;

Splist              : Subr_formal_para
                    | Splist TCOMMA Subr_formal_para
                    ;

Subr_formal_para    : Name
                    | TSTAR
                    ;

/*******************/

entry_statement : TENTRY Name
                | TENTRY Name TLPARN TRPARN
                | TENTRY Name TLPARN Splist TRPARN
                ;

/*******************/

block_data_statement: TBLOCK TZEOS
                    | TBLOCK Name TZEOS
                    ;

/*******************/

dimension_statement : TDIMEN Ardcllist ;

Ardcllist           : array_declarator
                    | Ardcllist TCOMMA array_declarator
                    ;

array_declarator    : Name TLPARN Ardims TRPARN ;

Ardims          : dim_bound_expr
                | TSTAR
                | dim_bound_expr TCOLON TSTAR
                | dim_bound_expr TCOLON dim_bound_expr
                | dim_bound_expr TCOMMA Ardims
                | dim_bound_expr TCOLON dim_bound_expr TCOMMA Ardims
                ;

dim_bound_expr  : arithmetic_expression ;

/*******************/
```

```
equivalence_statement   : TEQUIV Eqvsetlist ;

Eqvsetlist  : Eqvset
            | Eqvsetlist TCOMMA Eqvset
            ;

Eqvset      : TLPARN Eqv_entity_list TRPARN ;

Eqv_entity_list : equivalence_entity
                | Eqv_entity_list TCOMMA equivalence_entity
                ;

equivalence_entity  : Name
                    | array_element_name
                    | substring_name
                    ;

/******************/

common_statement: TCOMMO Common_block_list
                | TCOMMO Initial_blank_common Common_block_list
                | TCOMMO Initial_blank_common
                ;

Common_block_list   : Common_block
                    | Common_block_list Common_block
                    ;

Common_block    : TCNCAT Common_item_list
                | Common_block_name Common_item_list
                ;

Initial_blank_common: Common_item_list ;

Common_item_list: Common_item
                | Common_item_list TCOMMA Common_item
                | Common_item_list TCOMMA
                ;

Common_item     : Name
                | array_declarator
                ;

Optional_comma  : TCOMMA | /* empty */ ;

Common_block_name   : TSLASH TNAME TSLASH ;

/******************/

type_statement  : Data_type Type_elements
                | Character_type TCOMMA Type_elements
                ;
```

```
Character_type   : TCHARA TSTAR len_specification
                 ;

Type_elements    : Type_element
                 | Type_elements TCOMMA Type_element
                 ;

Type_element     : Type_item
                 | Type_item TSTAR len_specification
                 ;

Type_item    : Name
             | array_declarator
             ;

/*********** Implicit Statement***********/

implicit_statement  : TIMPLI Impdcllist
                    ;

Impdcllist       : Impdcl
                 | Impdcllist TCOMMA Impdcl
                 ;

Impdcl           : Data_type TLPARN Char_range_list TRPARN
                 ;

Char_range_list : Char_range
                | Char_range_list TCOMMA Char_range
                ;

Char_range       : Impchar TMINUS Impchar
                  Impchar
                 ;

/************** Datatypes and length specification **********/

Data_type        : TINTEG
                 | TINTEG TSTAR unsigned_int_constant
                 | TREAL
                 | TREAL TSTAR unsigned_int_constant
                 | TDOUBL     /* Double Precision */
                 | TCOMPL
                 | TCOMPL TSTAR unsigned_int_constant
                 | TLOGIC
                 | TLOGIC TSTAR unsigned_int_constant
                 | TDCMPL     /* Double Complex */
                 | TCHARA
                 | TCHARA TSTAR len_specification
                 ;
```

```
len_specification : unsigned_int_constant
                  | TLPARN Asterisk TRPARN
                  | TLPARN arithmetic_expression TRPARN
                  ;
```

/*******************/

```
parameter_statement : TPARAM TLPARN Para_decl_list TRPARN
                    ;

Para_decl_list      : Para_decl_item
                    | Para_decl_list TCOMMA Para_decl_item
                    ;

Para_decl_item      : Name TEQUAL expression
                    ;
```

/*******************/

```
external_statement  : TEXTER External_namelist
                    ;

External_namelist   : Name
                    | External_namelist TCOMMA Name
                    ;

intrinsic_statement : TINTRI Intrinsic_namelist
                    ;

Intrinsic_namelist  : Name
                    | Intrinsic_namelist TCOMMA Name
                    ;
```

/*******************/

```
save_statement  : TSAVE
                | TSAVE Save_item_list
                ;

Save_item_list  : Save_item
                | Save_item_list TCOMMA Save_item
                ;

Save_item       : Name
                | Common_block_name
                ;
```

/*******************/

```
data_statement   : TDATA Data_list ;

Data_list        : Data_declaration
                 | Data_list Optional_comma Data_declaration
                 ;

Data_declaration: Data_elements TSLASH Data_value_list TSLASH
                 ;

Data_elements    : Data_element
                 | Data_elements TCOMMA Data_element
                 ;

Data_value_list  : Data_value
                 | Data_value_list TCOMMA Data_value
                 ;

Data_element     : Data_var
                 | data_implied_do_list
                 ;

Data_var         : Name
                 | array_element_name
                 | substring_name
                 ;

Data_value       : Data_value_count TSTAR Single_data_value
                 | Single_data_value
                 ;

Data_value_count: unsigned_int_constant
                 | Name
                 ;

Single_data_value : Data_constant
                 | TPLUS unsigned_arithmetic_constant
                 | TMINUS unsigned_arithmetic_constant
                 | Name
                 ;

data_implied_do_list: TLPARN Data_impl_do_items TCOMMA Do_specification TRPARN
                 ;

Data_impl_do_items : Data_impl_do_item
                 | Data_impl_do_items TCOMMA Data_impl_do_item
                 ;

Data_impl_do_item : array_element_name
                 | data_implied_do_list
                 ;
```

```
Do_specification: Name TEQUAL arithmetic_expression TCOMMA arithmetic_expression
                | Name TEQUAL arithmetic_expression TCOMMA arithmetic_expression
                  TCOMMA arithmetic_expression
                ;

Data_constant    : unsigned_arithmetic_constant
                | character_constant
                | Hollerith_constant
                | logical_constant
                ;
```

/*******************/

```
assignment_statement: Name TEQUAL expression
                    | substring_name TEQUAL character_expression
                    | TASSIG Label_ass_ref TTO Name
                    ;

Assignment_or_stmt_fn : Asgn_or_sf_lhs TEQUAL expression ;

Asgn_or_sf_lhs      : Name TLPARN Arglist TRPARN ;
```

/*******************/

```
goto_statement  : TGOTO Label_cf_ref
                | TGOTO Label_list arithmetic_expression
                | TGOTO Label_list TCOMMA arithmetic_expression
                | TGOTO Name
                | TGOTO Name Label_list
                | TGOTO Name TCOMMA Label_list
                ;

Label_list      : TLPARN Labels TRPARN ;

Labels          : Label_cf_ref
                | Labels TCOMMA Label_cf_ref
                ;
```

/*******************/

```
arithmetic_if_stmt  : TIF TLPARN expression TRPARN Three_labels ;

Three_labels        : Label_cf_ref TCOMMA Label_cf_ref TCOMMA Label_cf_ref ;

logical_if_statement: TIF TLPARN expression TRPARN Statement_body ;

block_if_statement  : TIF TLPARN expression TRPARN TTHEN ;

else_if_statement   : TELSIF TLPARN expression TRPARN TTHEN ;

else_statement      : TELSE ;

end_if_statement    : TENDIF ;
```

```
/*******************/

do_statement      : TDO Label_DO_ref TCOMMA Do_specification
                  | TDO Label_DO_ref Do_specification
                  ;

/*******************/

continue_statement  : TCONTI ;

/*******************/

stop_statement   : TSTOP
                  | TSTOP Stop_or_pause_value
                  ;

/*******************/

pause_statement     : TPAUSE
                    | TPAUSE Stop_or_pause_value
                    ;

Stop_or_pause_value : unsigned_int_constant
                    | character_constant
                    ;

/*******************/

write_statement     : TWRITE control_info_list
                    | TWRITE control_info_list Outputlist
                    ;

control_info_list   : TLPARN unit_identifier TRPARN
                    | Unambiguous_cilist_2
                    ;

/*******************/

read_statement   : TREAD Simple_fmt_id
                 | TREAD Cilist_or_sp
                 | TREAD Unambiguous_cilist_1
                 | TREAD Simple_fmt_id TCOMMA Inputlist
                 | TREAD Cilist_or_sp TCOMMA Inputlist
                 | TREAD Read_cilist Inputlist
                 ;

/*******************/

print_statement : TPRINT format_identifier
                | TPRINT format_identifier TCOMMA Outputlist
                ;
```

```
Read_cilist       : Cilist_or_sp
                  | Unambiguous_cilist_1
                  ;

Cilist_or_sp      : TLPARN expression TRPARN
                  ;

Ci_item_list      : Control_info_item
                  | Ci_item_list TCOMMA Control_info_item
                  ;

Control_info_item  : TFMTKD TEQUAL format_identifier
                   | END_keyword TEQUAL Label_cf_ref
                   | ERR_keyword TEQUAL Label_cf_ref
                   | IO_keyword TEQUAL Ci_value
                   ;

format_identifier  : Label_io_ref
                   | Asterisk
                   | character_expression
                   ;

character_expression: Character_item
                    | character_expression TCNCAT Character_item
                    ;

Character_item    : Name
                  | character_constant
                  | Hollerith_constant
                  | Arelm_or_funref
                  | substring_name
                  | TLPARN character_expression TRPARN
                  ;

Ci_value          : expression | Asterisk ;

unit_identifier   : expression | Asterisk ;

Asterisk          : TSTAR ;

Simple_fmt_id     : Asterisk
                  | Label_io_ref
                  | SSimple_fmt_id
                  | SSimple_fmt_id TCNCAT character_expression

/* Cilist_or_sp used to prevent shift/reduce conflict */

                  | Cilist_or_sp TCNCAT character_expression
                  ;
```

```
SSimple_fmt_id  : character_constant
                | Name
                | Arelm_or_funref
                | substring_name
                ;

Unambiguous_cilist_1: Unambiguous_cilist_2
                    | TLPARN Asterisk TRPARN
                    ;

Unambiguous_cilist_2: TLPARN Ci_item_list TRPARN
                    | TLPARN unit_identifier TCOMMA Ci_item_list TRPARN
                    | TLPARN unit_identifier TCOMMA format_identifier TRPARN
                    | TLPARN unit_identifier TCOMMA format_identifier TCOMMA
                      Ci_item_list TRPARN
                    ;

Outputlist : Outputitem
           | Outputlist TCOMMA Outputitem
           ;

/* Note: This is all rather complicated due to potential conflicts
         during parsing with the complex constant */

Outputitem : expression
           | TLPARN Outputitem TCOMMA Outputlist TCOMMA Do_specification TRPARN
           | TLPARN Outputitem TCOMMA Do_specification TRPARN
           ;

Inputlist  : Inputitem
           | Inputlist TCOMMA Inputitem
           ;

Inputitem  : Input_var
           | TLPARN Inputlist TCOMMA Do_specification TRPARN
           ;

Input_var  : Name
           | array_element_name
           | substring_name
           ;

/******************/

open_statement  : TOPEN Auxiliary_cilist ;

/******************/

close_statement : TCLOSE Auxiliary_cilist ;

/******************/
```

```
inquire_statement   :TINQUI Auxiliary_cilist ;

Auxiliary_cilist    : TLPARN unit_identifier TRPARN
                    | TLPARN unit_identifier TCOMMA Aux_cispecs TRPARN
                    | TLPARN Aux_cispecs TRPARN
                    ;

Aux_cispecs         : Aux_cispec
                    | Aux_cispecs TCOMMA Aux_cispec
                    ;

Aux_cispec          : IO_keyword TEQUAL expression
                    | ERR_keyword TEQUAL Label_cf_ref
                    ;

/******************/

backspace_statement : TBACKS Fspec ;

/******************/

endfile_statement   : TENDFI Fspec ;

/******************/

rewind_statement : TREWIN Fspec ;

/******************/

Fspec               : unit_identifier
                    | TLPARN Asterisk TRPARN
                    | TLPARN Asterisk TCOMMA Aux_cispecs TRPARN

/* Silly outputitem in next production instead of expression because
   complex constant has to have it */

                    | TLPARN Outputitem TCOMMA Aux_cispecs TRPARN
                    | TLPARN Aux_cispecs TRPARN
                    ;

/******************/

format_statement : TFORMA TLPARN Fmtlist TRPARN
                 | TFORMA TLPARN TRPARN
                 ;

Fmtlist          : Fmtlist1  Fmtlist2 ;

Fmtlist1         : Fmtitem1
                 | Fmtitem1 TCOMMA Fmtlist
                 | Fmtitem1 Fmtlist2
                 ;
```

```
Fmtlist2          : Fmtitem2
                  | Fmtitem2 TCOMMA Fmtlist
                  | Fmtitem2 Fmtlist
                  ;

Fmtitem1          : Format_field
                  | Scale_factor
                  | Scale_factor Format_field
                  | character_constant
                  | Hollerith_constant
                  | Sub_format
                  | unsigned_int_constant Sub_format
                  ;

Fmtitem2          : TSLASH | TCOLON ;

Sub_format        : TLPARN Fmtlist TRPARN ;

Scale_factor      : TSCALE ;

/******************/

call_statement    : TCALL Name
                  | TCALL Name TLPARN TRPARN
                  | TCALL Name TLPARN Actualargslist TRPARN
                  ;

Call_argument     : expression
                  | TSTAR Label_cf_ref
                  ;

Actualargslist    : Call_argument
                    Actualargslist TCOMMA Call_argument
                  ;

/*******************/

return_statement    : TRETUR
                    | TRETUR arithmetic_expression
                    ;

/******************/

expression        : expression TEQV expression
                  | expression TNEQV expression
                  | expression TOR expression
                  | expression TAND expression
                  | TNOT expression
                  | Expr_2 TLT Expr_2
                  | Expr_2 TLE Expr_2
```

```
                | Expr_2 TEQ Expr_2
                | Expr_2 TNE Expr_2
                | Expr_2 TGT Expr_2
                | Expr_2 TGE Expr_2
                | logical_constant
                | Expr_2
                ;

Expr_2          : Expr_2 TCNCAT Expr_2
                | arithmetic_expression
                | substring_name
                | character_constant
                | Hollerith_constant
                ;

arithmetic_expression : Arithmetic_subexpr
                | TPLUS Arithmetic_subexpr
                | TMINUS Arithmetic_subexpr
                | arithmetic_expression TPLUS Arithmetic_subexpr
                | arithmetic_expression TMINUS Arithmetic_subexpr
                ;

Arithmetic_subexpr : Arithmetic_subexpr TSTAR Arithmetic_subexpr
                | Arithmetic_subexpr TSLASH Arithmetic_subexpr
                | Arithmetic_subexpr TDSTAR Arithmetic_subexpr
                | unsigned_arithmetic_constant
                | Name
                | Arelm_or_funref
                | TLPARN expression TRPARN
                ;

/*****************/

unsigned_arithmetic_constant: unsigned_int_constant
                    | unsigned_real_constant
                    | unsigned_dp_constant

/* The following two statements are both legal Fortran ...
    PRINT 5,(3,2)
    PRINT 5,(3,2,I=1,10)
  To get around the reduction conflict, complex constant accepts
  a rather revolting syntax as legal - this is horrible but basically
  unavoidable. */

                    | TLPARN Outputitem TCOMMA Complex_part TRPARN
                    ;
```

```
Complex_part       : Unsigned_part_of_complex
                   | TPLUS Unsigned_part_of_complex
                   | TMINUS Unsigned_part_of_complex
                   ;

Unsigned_part_of_complex: unsigned_int_constant
                        | unsigned_real_constant
                        | unsigned_dp_constant
                        ;

substring_name     : array_element_name Substring_spec
                   | Name Substring_spec
                   ;

array_element_name : Name TLPARN Arglist TRPARN ;

Arelm_or_funref : Name TLPARN Arglist TRPARN
                | Name TLPARN TRPARN
                ;

Arglist            : expression
                   | Arglist TCOMMA expression
                   ;

Substring_spec  : TLPARN Substring_element TCOLON Substring_element TRPARN
                ;

Substring_element  : arithmetic_expression | /* empty */ ;
/******************************************************************/
/*                                                              */
/* These productions are for tokens which have some value attached  */
/*                                                              */
/******************************************************************/

Name       : TNAME ;

IO_keyword : TNAME ;

ERR_keyword : TERRKD ;

END_keyword : TENDKD ;

Impchar    : TNAME ;

Label      : TDCNST ;

Label_cf_ref : Labelref ;

Label_DO_ref : Labelref ;

Label_io_ref : Labelref ;

Label_ass_ref : Labelref ;
```

```
Labelref     : TDCNST ;

unsigned_int_constant    : TDCNST ;

logical_constant     : TLCNST ;

unsigned_real_constant  : TRCNST ;

unsigned_dp_constant: TPCNST ;

character_constant   : TCCNST ;

Hollerith_constant   : THCNST ;

Format_field     : TFIELD ;

%%
```

Workshop on Fortran Analysis

Malcolm J. Cohen
NAG Ltd
Oxford, U.K.

"A periphrastic study in a worn-out poetical fashion,
Leaving one still with the intolerable wrestel
With words and meanings."

— Thomas S. Eliot,
East Coker, *in Four Quartets,* 1944

Abstract. This chapter describes the tools ISTVS, ISTVW, ISTVA and ISTAN. These tools are supplementary to the basic analysis facilities provided by Toolpack/1 Release 2.

Particular emphasis is given to the instrumentation facilities of ISTAN and to their integration into a Fortran environment based on Toolpack/1 tools.

1.   Introduction

This chapter describes the tools ISTVS, ISTVW, ISTVA and ISTAN, and shows some simple examples of their use. These tools supplement the basic analysis facilities of Toolpack/1 Release 2 described in the Chapter "Analysis Tools for Fortran 77".

2.   ISTVS – View Symbols

The Toolpack/1 symbol viewing tool, ISTVS, produces a formatted listing of the symbol table produced by the Toolpack/1 parser, ISTYP. The symbols for each program unit are listed separately, a section being output for each symbol type (label, common block, name, program unit identifier, variable, procedure, statement function and entry point) existing in the program unit. The symbols within each section are listed in alphabetic order.

A. A. Pollicini (ed.), Using Toolpack Software Tools, 101–119.
© 1989 ECSC, EEC, EAEC, Brussels and Luxembourg.

The listing is headed by user-supplied text concatenated with ': Symbol Table Listing' and followed by the date and time that the listing is being produced.

The output from this tool is not intended to be suitable for inclusion "as is" in documentation. In particular, the output includes such things as parse tree node numbers, which are only useful in connection with the parse tree. Instead, this tool is meant to be a low-level facility for examining symbol tables.

Example VS-1 shows some typical output from this tool.

Example VS-1

Input program:

```
PROGRAM VS1
CALL B(4.)
END
SUBROUTINE B(Y)
REAL X
PRINT *,TAN(Y**2)
END
```

Output from ISTVS:

```
Example VS-1: Symbol Table Listing, 13:00:30 12 NOV 1986

Program Unit: VS1 Main Program
        Procedures:
            B SUBROUTINE
                Called as a subroutine
Program Unit: B SUBROUTINE
        Names (Usage Unknown):
            X REAL
                Explicitly typed
        Variables:
            Y REAL
                Formal parameter
                In an expression
        Procedures:
            TAN Generic
                Standard intrinsic function
                Called as a function
                In an expression
```

3. ISTVW – View Warnings

ISTVW analyses the symbol table produced by the Toolpack/1 parser, ISTYP, to produce a formatted listing of warning messages.

The listing is headed by user-supplied text concatenated with ': Simple Warnings Listing' and followed by the date and time that the listing is being produced.

The warning summary contains information about some common features of a program unit that often (but by no means always) imply that something has been overlooked. The warnings which may be given in Toolpack/1 Release 2 include:

(1) A variable, statement function, parameter or external procedure has been implicitly typed.

(2) A variable, symbol, dummy argument, statement function, parameter or procedure has been declared but not used.

(3) A local variable has been declared but not explicitly assigned a value in this routine.

(4) A label has not been used.

(5) A function does not have its value set.

(6) A symbol name is not legal. A report is given if a name does not conform to the Fortran 77 standard or to the local compiler (as defined in ZLEGAL in the String Supplementary Library) or to both.

(7) A standard intrinsic function has been explicitly typed or does not appear in an INTRINSIC statement.

(8) An external procedure does not appear in an EXTERNAL statement.

Example VW-1 gives some typical output from this tool.

Example VW-1

Output from ISTVW:

```
Example VW-1: Simple Warnings Listing, 13:00:42 12 NOV 1986

Program Unit: VS1 (Main Program)
  External procedure not in EXTERNAL: B

Program Unit: B (Routine)
  Unused symbol: X
  Implicitly typed Variable: Y (REAL)
  Intrinsic procedure not in INTRINSIC: TAN
```

4.   ISTVA – View Attributes

ISTVA produces a report from the information contained in the modified symbol table and attribute file created by the Toolpack/1 static semantic analyser, ISTSA. The output from ISTVA is intended to be generally usable as documentation, unlike that from the symbol viewing tool, ISTVS.

4.1.   The Extended Symbol Table

The first part of the report produced by ISTVA is from the extended (local) symbol table. The local symbol tables contain program-unit specific information only.

The information listed about each name occurring within a program-unit is a superset of that listed by ISTVS. In particular, array sizes and dimensions are given, and the following additional status information is listed when it has been discovered by ISTSA:

(1) Actual argument to external,

(2) Parameter value known,

(3) Equivalenced into a common block,

(4) In an array declarator.

If the variable is in a COMMON block then the COMMON block name is given together with the offset within the block. If the variable is in a local equivalence class then this class is identified (by a small integer) and its relative storage location within that class is given. In either case the storage location or offset is given in character storage units.

In the listing of local common blocks, the variables occurring in common statements are listed, together with the total size (in character storage units) and usage of the common block (the usage is the inclusive-or of the status bits of all variables in the common block).

4.2.   The Global Symbol Table

The second part of the report produced by ISTVA is the global symbol table. This report is divided into three parts: program-units, common-blocks and unsatisfied external references.

The program-unit section lists the arguments, common blocks and external references of each program-unit. The argument information includes whether the arguments are updated, read-only or passed out to another routine (and so possibly updated). The common block information consists of the names of all common blocks and whether each common block is updated.

The common block section lists every common block in the program with its size, whether it is SAVEd or not, and whether it consists of character data, numeric data, or both.

The unsatisfied external reference section lists external references to routines which do not occur in that part of the program given to ISTSA, and also "indirect" external references (i.e. to procedure arguments). The types of these routines and their arguments are also listed.

Example VA-1

Input program:

```
          PROGRAM EXAMPL
          COMMON /A/ X,Y
          REAL B,C,FUN,X,Y
          EXTERNAL FUN
          B=1.5
          X=2.
          Y=3.
          C=FUN(B)
          WRITE (*,100) C
100       FORMAT (' ',F10.3)
          END

          REAL FUNCTION FUN(P)
          COMMON /A/ B,C
          REAL P,B,C,Y,Z
          EXTERNAL N12345
          Y=B*C
          CALL N12345(Y,Z)
          FUN=Z/P
          END
```

The two parts of the report output from ISTVA are headed by a user-supplied text, followed by the table name and by the date and time stamp.

The name of the specific symbol table, either local or global, is represented by the string ': Extended Symbol Table Listing' for the former part of the report, and by ': Global Attribute Listing' for the latter.

Remark that the appropriate header line is repeated at the beginning of each page of the listing.

Example VA-1: Extended Symbol Table Listing, 10:18:29 21 AUG 1986

Program Unit: EXAMPL PROGRAM
        Labels:
            100, Parse tree node 42
                Referenced by 1 I/O statements (as FORMAT)
        Common blocks:
            A, Size: 8
                Items: X, Y
                Usage:
                    Explicitly typed
                    Assigned to on left of "="
                    In COMMON statement
        Variables:
            B REAL
                Explicitly typed
                Assigned to on left of "="
                Used as an actual argument
                In an expression
                Actual argument to external
            C REAL
                Explicitly typed
                Assigned to on left of "="
                In an expression
            X REAL
                Explicitly typed
                Assigned to on left of "="
                In COMMON statement
                In common block /A/, offset 0
            Y REAL
                Explicitly typed
                Assigned to on left of "="
                In COMMON statement
                In common block /A/, offset 4
        Procedures:
            FUN REAL
                Declared EXTERNAL
                Explicitly typed
                Called as a function
                In an expression

Program Unit: FUN REAL FUNCTION
                Explicitly typed
                Assigned to on left of "="

Example VA-1: Extended Symbol Table Listing, 10:18:29 21 AUG 1986

        Common blocks:
            A, Size: 8
                Items: B, C
                Usage:
                    Explicitly typed
                    In COMMON statement
                    In an expression
        Variables:
            B REAL
                Explicitly typed
                In COMMON statement
                In an expression
                In common block /A/, offset 0
            C REAL
                Explicitly typed
                In COMMON statement
                In an expression
                In common block /A/, offset 4
            P REAL
                Formal parameter
                Explicitly typed
                In an expression
            Y REAL
                Explicitly typed
                Assigned to on left of "="
                Used as an actual argument
                In an expression
                Actual argument to external
            Z REAL
                Explicitly typed
                Used as an actual argument
                In an expression
                Actual argument to external
        Procedures:
            N12345 SUBROUTINE
                Declared EXTERNAL
                Called as a subroutine

Example VA-1: Global Attribute Listing, 10:18:29 21 AUG 1986

Program Units
=============
    EXAMPL: PROGRAM
        Common Block /A/, updated
        Calls FUN

    FUN: REAL Function
        Argument 1: REAL , read-only
        Common Block /A/
        Calls N12345

Common Blocks
=============
    /A/, Length 8, non-character

External References
==================
    N12345: SUBROUTINE
        Argument 1: REAL Scalar
        Argument 2: REAL Scalar

5.   ISTAN – The Execution Analysis Tool

ISTAN is a tool which *instruments* a Fortran 77 program; i.e., from the user's program it produces another program which has the same effect but has *instrumentation* code added to it. In the case of ISTAN, this instrumentation code may:

(1) perform execution frequency counting of code segments,

(2) count assertion failures (assertions being special comments containing Fortran 77 logical expressions which the programmer expects to be true at that point),

(3) perform trace analysis: this may be

    (a) execution tracing of particular code segments,

    (b) control flow trace following execution of particular code segments, or

    (c) control flow trace prior to execution of particular code segments.

(4) update history files containing cumulative frequency counts over a number of program executions.

Having processed their program using ISTAN, the users must then compile the instrumented code produced by ISTAN using the host Fortran system to produce an instrumented version of their executable program. This instrumented program may then be executed to produce the data required by the users.

ISTAN was developed from the Fortran 77 Analyser [48] developed for the National Bureau of Standards.

5.1.   Program Segments

Central to the operation of ISTAN is the concept of a segment. Basically, a segment is a "straight-line" section of Fortran code, i.e. one which may only be entered at one position (the top) and exits at one position (the bottom) with each statement in the segment being executed exactly once each time the segment is entered.

5.2.   The Instrumentor

ISTAN takes as its input a Fortran program in token stream form (from the Toolpack/1 lexical analyser, ISTLX) and produces an instrumented Fortran source, a statement summary file for input to ISTAL [37], an annotated token stream and a summary report.

The input program may either be TIE-conforming or normal Fortran. Only standard conforming Fortran 77 [1] or Toolpack/1 target Fortran [13] programs can be instrumented. The examples in this section are all based on the program shown in example AN-1.

Example AN-1

```
      PROGRAM TEST
      INTEGER X(10),NEVEN,SUM,RESULT,ND3
      DATA X/1,2,3,4,5,6,7,8,9,10/
      PRINT *,'Of the integers from 1 to 10:'
      RESULT=ND3(X,10)
      IF (RESULT.EQ.0) THEN
          PRINT *,'None are divisible by 3'
      ELSE IF (RESULT.EQ.1) THEN
          PRINT *,'One is divisible by 3'
      ELSE
          PRINT *,RESULT,' are divisible by 3.'
      END IF
      END
      INTEGER FUNCTION ND3(A,N)
      INTEGER N,A(N),I
 *$as$(N.GT.0)
      ND3=0
      DO 100 I=1,N
100        IF (MOD(A(I),3).EQ.0) ND3=ND3+1
 *$as$(ND3.GE.0)
      END
```

The annotated token stream lists the segment numbers (a segment is a section of straight-line code) used in later reports: these are in the form of source-embedded directives (also described in [13]).

The input may contain assertions in the form of source-embedded directives, which may optionally be inserted into the instrumented code as runtime checks.

It is not necessary for the input to contain a complete Fortran program. If only a few routines are to be analysed, they may be input to ISTAN and the instrumented output combined with the rest of the program.

5.3.   The Annotated Token Stream

The annotated token stream should be polished (by ISTPL) to produce the annotated listing. This still has the segment numbers appearing as source-embedded directives; to produce a listing more suitable for documentation purposes it may be processed by ISTAL. Example AN-2 shows the annotated listing produced by the sequence ISTAN, ISTPL and ISTAL from the program in example AN-1.

Example AN-2

```
SEGMENT 1:
            PROGRAM TEST
            INTEGER X(10),NEVEN,SUM,RESULT,ND3
            DATA X/1,2,3,4,5,6,7,8,9,10/
            PRINT *,'Of the integers from 1 to 10:'
            RESULT = ND3(X,10)
            IF (RESULT.EQ.0) THEN
SEGMENT 2:
                PRINT *,'None are divisible by 3'
SEGMENT 3:
            ELSE IF (RESULT.EQ.1) THEN
SEGMENT 4:
                PRINT *,'One is divisible by 3'
SEGMENT 5:
            ELSE
                PRINT *,RESULT,' are divisible by 3.'
SEGMENT 6:
            END IF
            END
```

```
SEGMENT 7:
              INTEGER FUNCTION ND3(A,N)
              INTEGER N,A(N),I
ASSERTION 1:
       *(N.GT.0)
              ND3 = 0
              DO 100 I = 1,N
SEGMENT 8:
SEGMENT 9:
          100 IF (MOD(A(I),3).EQ.0) ND3 = ND3 + 1
ASSERTION 2:
       *(ND3.GE.0)
SEGMENT 10:
              END
```

5.4.   The Summary Report

The summary report file produced by ISTAN contains a formatted summary of the statement types found in the input program. It also contains counts of:

(1) Assertions processed (zero if the assertions option is not specified).

(2) Comment lines.

(3) Errors detected.

(4) Tokens input (excluding comments).

(5) Statements.

Note that the conditional part of a logical IF is not counted as a statement either in the total or in the statement type counts.

Example AN-3 shows a sample summary report.

Example AN-3

STATEMENT TYPE SUMMARY

ASSERTIONS	2
COMMENTS	0
ERRORS	0
TOKENS	150
STATEMENTS	19

ASSIGN	0	GO TO	0
BACKSPACE	0	(ASSIGNED)	0
BLOCK DATA	0	(COMPUTED)	0
CALL	0	(UNCONDITIONAL)	0

CHARACTER	0	IF	2
CLOSE	0	(ARITHMETIC)	0
COMMON	0	(BLOCK)	1
COMPLEX	0	(LOGICAL)	1
CONTINUE	0	IMPLICIT	0
DATA	1	INQUIRE	0
DIMENSION	0	INTEGER	2
DOUBLE PRECISION	0	INTRINSIC	0
DO	1	LOGICAL	0
ELSE IF	1	OPEN	0
ELSE	1	PARAMETER	0
ENDFILE	0	PAUSE	0
END IF	1	PRINT	4
END	2	PROGRAM	1
ENTRY	0	READ	0
EQUIVALENCE	0	REAL	0
EXTERNAL	0	RETURN	0
FORMAT	0	REWIND	0
FUNCTION	1	SAVE	0
CHARACTER	0	STOP	0
COMPLEX	0	SUBROUTINE	0
DOUBLE PRECISION	0	WRITE	0
INTEGER	1	(ASSIGNMENT STATEMENTS)	2
LOGICAL	0	(STATEMENT FUNCTIONS)	0
REAL	0	(UNRECOGNIZED STATEMENTS)	0
UNTYPED	0	-	0

5.5.   The Instrumented Program

The instrumented program produces as output a listing file, an optional history file, optional tracing information and an optional single-run data file. These are in addition to any output normally produced by the (non-instrumented) program.

- **The Listing File**

This file contains a formatted report of the execution frequencies of each segment in the instrumented program and a list of all segments which were not executed. If assertion checking was requested the report will additionally contain the results of the assertion checks. Example AN-4 is the program output plus the listing file generated from the execution of the program from example AN-1.

Example AN-4

```
Of the integers from 1 to 10:
3 are divisible by 3.
```

	0	1	2	3	4	5	6	7	8	9
\multicolumn{11}{c}{SEGMENT EXECUTION FREQUENCIES - CURRENT}
TEST										
OX		1	0	1	0	1	1			
ND3										
OX								1	10	3
1X	1									

SEGMENTS NOT EXECUTED

```
2    4
```

	0	1	2	3	4	5	6	7	8	9
\multicolumn{11}{c}{ASSERTION FAILURE FREQUENCIES - CURRENT}
ND3										
OX		0	0							

- **History**

Each time the instrumented program is run the frequency counts for that execution will be added to the data from the input history file to produce the output history file (this may be the same file). The history file may later be analysed by ISTAL to produce a cumulative report. (History file processing only occurs if the history option is specified to ISTAN).

- **Trace Information**

When the instrumented program is run, it will read the trace range information from the trace input file. Trace information will then be gathered during execution and periodically written to the trace output file. (Trace processing is only performed if the trace option is specified to ISTAN).

- **Single-Run Data**

The single-run data file is written by the instrumented program upon termination and contains the segment execution frequencies (and possibly assertion check information) for that run in a form suitable for input to ISTAL. (The single-run data file is only produced if the rundata option is specified to ISTAN).

5.6.   Integration into Toolpack/1

ISTAN is integrated with three other Toolpack/1 software tools: the lexical analyser
ISTLX, the polisher ISTPL and the documentation generation aid ISTAL. The diagram
of Figure 1 shows the flow of data between these tools and the local Fortran system.

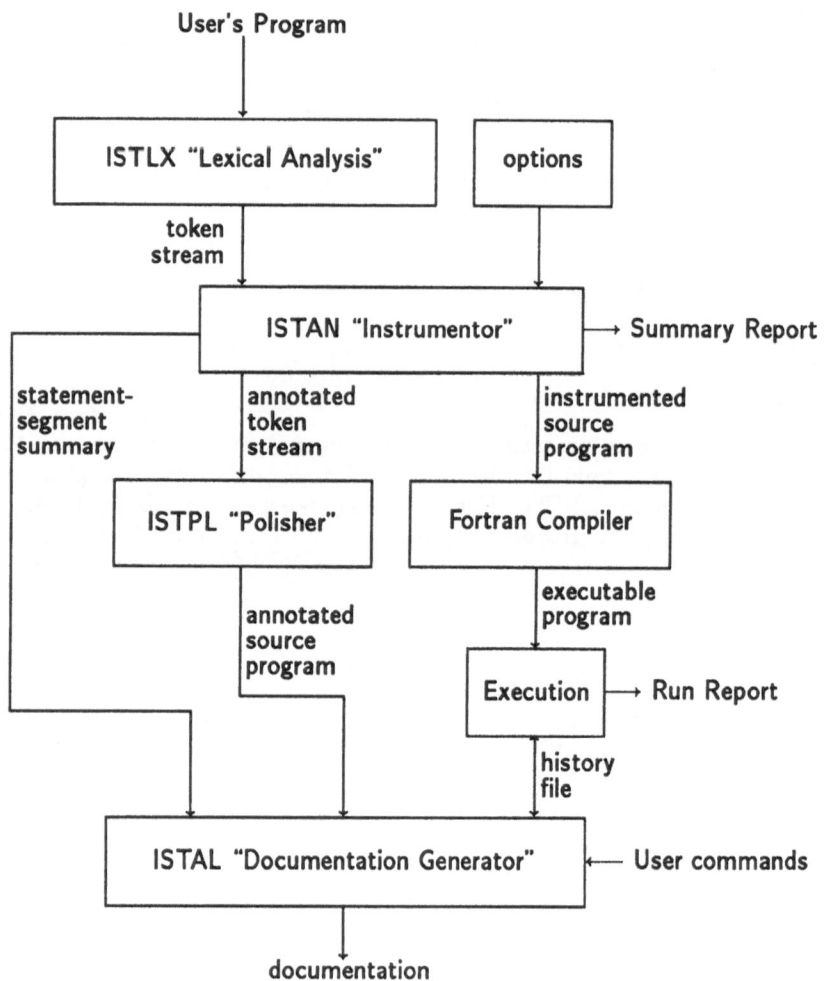

Figure 1. ISTAN: Integration into Toolpack/1

5.7. Options

The options available for ISTAN are listed below. Any unambiguous abbreviation for an option is acceptable. The first four characters of any option will always be unique, even in later releases of ISTAN.

assertions

This option requests assertion monitoring. Assertion comments in the input source code are compiled into run-time checks in the instrumented code. If there are no assertion comments in the input this option has no effect. If this option is not specified then assertion comments will be ignored.

history=arg1,arg2

This option requests history file processing by the instrumented program. If this option is not specified then no history information will be produced. The first argument specifies the input history file, and the second specifies the output history file. If the second argument is not included, then the updated history will be written either:

(a) For TIE-conforming Fortran code, to the same file as the input history was read from, otherwise

(b) to a site-dependent unit number (default 8). Unless the history input unit number has been changed from the default, this will result in the overwriting of the input history file.

Possible argument formats are:

(1) An asterisk. This results in the actual filename to be used being requested from the user at runtime.

(2) A filename in single quotes. This is assumed to be a host file name unless the tie_conforming option is used.

(3) An integer. This is the unit number to be used (normal Fortran) or the pre-connected device number (TIE-conforming Fortran). In the latter case note that the pre-connected device numbers are 0 to 3 only. This option may be used to mix program I/O and instrumentation I/O (e.g. trace output).

(4) An integer and a filename, e.g. " history=3='fred.hst' ". This format may not be used when instrumenting TIE-conforming Fortran, and specifies both the unit number to be used and the name of the file to be opened. This may be used to avoid unit number clashes between the instrumented program's I/O and the instrumentation I/O.

listing=argument

This option specifies where to output the listing file from the instrumented program. This file is always produced, and by default goes to the standard output device. The argument has the same format as the history option arguments.

rundata=argument

This option requests the production of a single-run execution data file. The argument has the same format as the history option arguments.

tie_conforming

This option specifies that the input source code is TIE-conforming, and so the instrumented program must use the TIE-conforming runtime library. If this option is not specified, the input is assumed to be normal (i.e. not TIE-conforming) Fortran.

trace=arg1,arg2

This option requests the production of tracing information by the instrumented program. The arguments have the same format as the history option arguments, and define the trace range input file and the trace information output file respectively. The default values for the arguments specify that the trace ranges are to be input from the standard input device (e.g. terminal) and the trace information is to be output to the standard output device (e.g. terminal). If this option is not specified then no tracing will be done.

vname=argument

This option determines the last 5 characters of all variables and routines added by ISTAN to produce the instrumented program. If this option is not specified, then these names will end with "ZZ4QX".

5.8.   Assertion Processing

Assertion processing is requested by the assertions option. An assertion is a source-embedded directive in the input program with an id of "as" and a logical expression inside parentheses as its body.

Example:

```
*$as$ (I.GE.1 .AND. I.LE.10)
```

An assertion may be continued on successive lines, and will be terminated by a closing parenthesis. Thus,

```
*$as$ ((I.NE.0) .OR.
*$as$ FOUND)
```

is a two-line assertion, while

```
*$as$ (I.NE.0)
*$as$ (FOUND)
```

are two separate assertions.

During execution of the instrumented program, the assertions are checked when encountered, and if false then the assertion failure is counted and reported at the end of the program. Note that since assertions are translated by the instrumentor into statements to perform the run-time checks, they must appear only in places where executable statements may legally appear (e.g. not between continuation lines of another statement).

5.9. History File Processing

History file processing is requested by the history option. If this option is not specified no history-processing code is inserted into the instrumented program. History information consists of segment execution frequencies and (if assertion processing is being done) assertion failure frequencies.

When the program terminates the history file is read, the data for this execution are added to the history data and a new history file is written.

If no history file is present the history output file will contain only the data for the current execution.

5.10. Segment Tracing

Segment tracing is requested by the trace option. If this option is not specified no segment tracing code is inserted into the instrumented program.

When the program is run, trace range requests will be read from the trace input file (and displayed on the trace output file). Trace range requests are input in the form "*IIIIsJJJJ*" where *IIII* and *JJJJ* are right-justified integers and *s* is a comma, plus sign or minus sign (if the tie-conforming option is used trace input is free-format).

Possible trace range requests are:

(a) *IIII,JJJJ*
Trace code segments *IIII* to *JJJJ* inclusive (range tracing).

(b) *IIII+JJJJ*
Trace code segment *IIII* and the next *JJJJ* segments to be executed (post-dated tracing).

(c) *IIII-JJJJ*
Trace code segment *IIII* and the *JJJJ* segments which were executed prior to segment *IIII* (pre-dated tracing).

If no trace ranges are input then no tracing will be done. Trace ranges may be input in any order and may overlap. There is a site-dependent limit (normally 25) on the number of separate trace range requests which may be entered.

Trace statistics are collected in a buffer which is periodically output to the trace output file. Data produced by range tracing and post-dated tracing are output with each line beginning with "TRACE   =". Pre-dated tracing data is output with "TRACE(PRE)=" at the beginning of each line. Repetition counts are used to reduce the size of the output. A repetition is indicated by a minus sign, thus "10 -7" would indicate that segment 10 was executed 7 times in succession. Example AN-5 shows the trace output produced from the program appearing in example AN-1.

To interpret the results of tracing, the trace output is used in conjunction with the annotated listing. Cross-referencing examples AN-2 and AN-5, we can see that:

segment 1   (the first part of the main program) was executed, then

segment 7   (the first part of the function ND3) was executed,

segment 8   (the logical IF test within the DO loop) was executed 3 times in succession,

segment 9   (the conditional statement of the logical IF) was executed,

　　　... and so on.

Example AN-5

```
TRACE OUTPUT REQUESTS
  TRACE=   1,  10

TRACE=        1   7   8   -3   9   8   -3   9   8   -2   8
TRACE=        9   8  10    3   5   6
```

5.11.   Single-Run Statistics

The single-run data file will be produced if the **rundata** option was specified to ISTAN. It has exactly the same format as the history file, but contains data for the current execution only.

5.12.   Partial Instrumentation

ISTAN may be used to instrument only a part of the user's program. This part must consist of complete subprograms (i.e. you cannot instrument only part of a subprogram).

To partially instrument a program, simply split it into two or more files, one of which contains exactly those routines which require instrumentation. This file should then be processed in the usual manner by ISTAN to produce the instrumented version of those routines.

If the part of the program which is being instrumented does not include any STOP statements or the main program unit, ISTAN will issue a warning saying that the user must insert calls to the "wrapup" subroutine at program termination points. This subroutine has no parameters, and its name is 'R' followed by the value of the **vname** option (usually 'ZZ4QX'). The wrapup routine terminates with a STOP statement itself, so the user simply replaces "STOP" with, for example, "CALL RZZ4QX". This must be done to all program termination points, even if ISTAN doesn't produce its warning (i.e. because it found at least one termination point), if the instrumentation is to function correctly.

For TIE-conforming programs (which call ERROR or ZQUIT instead of using STOP or the main program END), calls to ERROR should be replaced by 'E' followed by the **vname** option value and calls to ZQUIT by 'R' followed by the **vname** option value, e.g. "CALL ZQUIT(ok)" becomes "CALL RZZ4QX(ok)" and "CALL ERROR('MSG')" becomes "CALL EZZ4QX('MSG')".

5.13.   Restrictions

There are few limitations to the instrumentation capabilities of ISTAN which have to be taken into account:

(1) Tracing of subroutines and functions invoked in an I/O statement will fail on some compilers (as it may cause a recursive I/O call).
(2) In the statement summary, DOUBLE COMPLEX statements and functions will be listed as COMPLEX statements and functions. Forms such as "REAL*8" will be listed under their keyword, in this case as "REAL".
(3) The routine ZEXIT cannot be successfully instrumented, thus ISTAN cannot be used to instrument a TIE-conforming command interpreter (e.g. ISTCE).

The Fortran 77 Source Polisher

Aurelio A. Pollicini
Joint Research Centre
Ispra, Italy

"Ordine, chiarezza, semplicità!"

— Giosuè Carducci,
recommendation addressed to the candidates for a
University admission competition, as reported in
G. Pascoli, Ricordi di un vecchio scolaro,
"Il Resto del Carlino", 9th Feb. 1896

Abstract. The general aspects of the presentation of Fortran 77 source code in a tidy and polished form are discussed. Then, the way in which the polishing mechanism is implemented in the Toolpack environment is described, putting particular emphasis on the user interfaces of the tools ISTPO and ISTPL which provide the facility.

1.   A Dissertation on Presentation Rules

Programming languages should not be exclusively seen as means for "talking" to a computer, but also for exchanging formal ideas among people. Thus a computer program not only specifies actions a processor is expected to perform, but also describes a formal instance of some process devised by human imagination for the usefulness of a community of individuals. What more beneficial goal is achieved if that computer program, in addition to be correct, reliable and efficient, is also easy to understand as well as provided with an attractive look. That is matter of style. The elements of a sound programming style may be gathered in different main areas: structured design, appropriate choice of features and techniques and aesthetic presentation. We are not committed here to discuss structure and components of a program. We assume that the program is hierarchically structured in small, self-contained modules; each one clearly mapped on a well-defined functionality and all of them interfacing correctly. We also assume that it specifies consistently the attributes of the defined data-objects and it

121

A. A. Pollicini (ed.), Using Toolpack Software Tools, 121–130.
© 1989 ECSC, EEC, EAEC, Brussels and Luxembourg.

operates on those objects unambiguous transformations in the frame of orderly nested constructs.

Our purpose is to talk about presentation.

As in the Universe, the target of our aesthetic exercise is to reach order and harmony in the source program. What are the conceptual means at our disposal? Let us consider some philosophical principles and then try to implement the predicates implied by them.

Clarity makes the distinction between unity and diversification

Simplicity is the basis of harmonious complexity

Discipline is the foundation of ordered behaviour

From similar principles one may establish some basic rules for the presentation of the code. Obviously, such rules change from language to language and we confirm, of course, our attention to Fortran 77. When the same purpose may be achieved in different ways:

(1) Clarity requires that the more explicit one, in term of visibility of attributes and properties involved, must be preferred. For instance, code:

```
REAL        PRESS(100)
```

rather than:

```
DIMENSION PRESS(100)
```

(2) Simplicity suggests you to choose the less verbose form. For instance, code:

```
INTEGER    ATOMS(92)
```

rather than:

```
DIMENSION ATOMS(92)
INTEGER    ATOMS
```

(3) Discipline encourages you to use the same form throughout the entire program.

Following these principles a comprehensive set of presentation rules can be synthesized which provide the backbone of a recommended style. An exercize of this sort is shown in the Chapter "Supporting Coding Conventions".

2.   Approaches to Tidying Source Programs

Once a given set of presentation rules has been agreed upon, we are concerned with how to implement them. And still more important, we have to decide whether to apply them limitedly to newly developed software or to rearrange existing software according to them. The former situation is typical of a software development environment. Any systematic approach to software development aims, by various means, at the production of readable code. Coding guidelines were the first help for disciplined programmers. However, as any manual means, it was neither easy nor common to have them applied consistently. In practice, they proved to be inadequate if not enforced by an automatic tool. Indeed, tidying utilities were probably the first class of automatic programming aids for supporting the development of software. During the development process, software either grows stepwise, or evolves by prototyping. In particular, iteration over the phases of design, coding and testing causes different versions of the software to be compiled and run subsequently. The same principles used for the reduction of algorithms would suggest to move automatic tidying out of the iteration since the transformations applied to each version are independent of any intermediate version previously transformed. Thus the expected effect of presenting the final product in a nice and tidy form may be achieved effectively, by polishing the latest version only. However, it is useful to work on readable and understandable code throughout the whole development process. Therefore, the availability of an automatic tool makes it easy to have a uniform style for all developing versions. Moreover, we must be aware of the need for continual verification that automatic transformations have preserved computational equivalence, thus avoiding any risk of *over-reliance*, as mentioned in the general considerations in Section 3.2 of the Chapter "The Toolpack Project". In practice, the safer approach tends to polish every source code before compiling it; though, with the increasing confidence in the tool, it may be sensible to polish extensively modified versions only.

The latter situation is more tied to the operation of application software in regular use. Polishing these products may have the double benefit of generalizing the presentation style and of improving understandability of the software in view of possible future requirements for adaptation/maintenance. The application of an automatic tool, in this case, is mandatory.

Summing up, we have seen that the availability of an automatic tool is necessary for providing a solution to both aspects of the problem, namely presentation of new and existing software. We will now consider how such a goal can be reached by using the polishing mechanism provided by Toolpack/1 software.

3.   Structure of the Polishing Mechanism

According to the Toolpack policy of tool fragmentation for achieving a greater flexibility, the action of tidying the source view of a Fortran 77 program is not obtained by a single tool. The actual source transformation is performed on an already scanned form of the program, that is the view produced by the lexer ISTLX in term of a stream of lexical tokens as well as a stream of comments. That fundamental preliminary action on any source text has already been described in the Chapter "Analysis Tools for Fortran 77". We will assume that lexical analysis has previously been performed, outputting both files in the environment we are using. (Either host files or files in the Portable File Store PFS). The tidying function requires the use of one tool: the polisher ISTPL which returns back to the user a computationally equivalent source program whose formal presentation fulfils the requirements of a set of presentation rules. A second tool completes the polishing mechanism provided by Toolpack/1. It is the polish options setting tool ISTPO. The task performed by this tool is to establish a parametric view of the actual set of presentation rules in the form of a file. All parameters have a mnemonic keyword which is unique within the file and a pre-defined default value. The tool ISTPO allows for the interactive editing of the option file so that any parameter may be assigned a user-specified value which overwrites the default in the option file.

The following two sections will describe the use of ISTPL and ISTPO respectively.

4.   ISTPL User's Interface

Following the general rule for Toolpack/1 tools, ISTPL has a number of external interfaces which must be made available from the environment; if those interfaces are not implicitly provided in some automatic way such as the command executor ISTCE or the command interpreter ISTCI, the user is prompted for them. In the case of ISTPL there are three input files and one output file which the tool expects in the following order:

(1) token stream (the first input);

(2) comment stream (the second input);

(3) polished source (the output);

(4) polish options (the third input).

In addition to the four interfaces specified above, ISTPL is connected to a primary user input (standard input file) and two primary outputs to the user (standard output file and standard error file). In a conversational environment the standard input file is the user keyboard, while both standard output and standard error files are displayed at the output device of the user terminal; e.g. the screen of a video-terminal or any other equivalent means.

Once the names of the four files previously mentioned are known to ISTPL, the tool initiates to process the lexical view of the source program to be transformed. At completion of the transformations implied by the actual values of the option file, one of the two following messages is issued to the user terminal:

    ISTPL Normal Termination

or:

    ISTPL Termination, n Errors Found

In the latter case one or more error messages had appeared at the user terminal before the completion message that reports the total number "n" of errors.

The remaining part of this section will give details of some of the most common errors, while the full list of Error Messages is reported as an appendix of the *"ISTPL/ISTPO Users' Guide"*[8].

We may classify the errors according to their different scopes.

4.1.   Failure during Physical File Access

The messages belonging to this class correspond to fatal errors. They are:

- ''Can't Open token path''
- ''Can't Open Comment path''
- ''Can't Open option file''

These three messages mean that the corresponding file does not exist. Check the spelling of the file name and, if correct, verify whether the file was actually created by some previous run or not.

- ''Can't Open output file''

When the polished source is to be output to an existing host file, the open operation with intent of overwriting the file may be impeded in some rigid or over-protected host environments. Either delete it or change name.

- ''Can't create temporary scratch file''

This message corresponds to an internal table overflow. However, the limit is set to 999 so that the error should not occur for normal use.

- ''Attempt to Read Past End-of-File''

Some input file does not end correctly.

4.2.   Ill-conditioned Scanned View

This class of messages consists of two groups. The former refers to fatal errors caused either by invalid tokens or by label inconsistencies. The messages are:

- ''Incomplete token file''
- ''Token Read Failed''
- ''Token too long, recovery impossible''

For various reasons, the used token file was not generated correctly.

- ''Undefined labels in Program Unit''
- ''Duplicate labels''
- ''PROBL called with label=0''
- ''SETLBL - Internal Error''
- ''SETLBL - Catastrophic Error''

Defined and referenced labels do not correspond correctly.

The following messages, on the contrary, refer to situations that the polishing tool will try to recover:

- ''Missing END statement''
- ''DO nesting level>0 at END of Program Unit''
- ''IF nesting level>0 at END of Program Unit''
- ''Unlabeled FORMAT statement''
- ''Label too big for requested label column''
- ''Unbalanced parentheses''

4.3.   Option Inconsistencies

All the messages below correspond to fatal errors:

- ''RMARGC>72 and sequence number requested''

Either set the option controlling the right margin of comments to 72 or turn off the flag SEQRQD.

- ''DOCONI And Not RLBSTM''

The request to end each DO-loop with a CONTINUE statement implies generation of new labels in the case of nested loops sharing the same termination statement. Therefore, turn on the flag RLBSTM.

- ''INDDOC And Not DOCONI''

Turn on the flag DOCONI if you wish that CONTINUE statements at the end of DO-loops be indented as the loop body. Indeed, the option INDDOC requires all DO-loops to terminate on a CONTINUE statement.

- ''LMARGS is greater than RMARGS''

Set both options consistently.

4.4.   Implementation Restrictions

There are two specific situations that may be overcome by reinstalling the polisher ISTPL, with different customization parameters. They are pointed out by the following messages:

- ''DO-loops nested too deeply''
- ''Too many labels''

The specific internal limits are exceeded, therefore the tool is unable to transform the source program being processed. However, it is possible to extend the polisher capabilites. For doing that, assign larger values to the symbolic macros "maxDOlevel" (default is 30) and "maxlabels" (default is 500) to be found in the file PLDEFS, and then reinstall a new version of the tool.

5.   ISTPO User's Interface

ISTPO interfaces to one file only. Such a file is used to store the options that implement the set of presentation rules ISTPL will obey, in order to produce a polished version of the original version of a user program. Standard input, standard output and standard error files are also available. However, the most interesting part of the user interface is at the lower level structure of the option file which may be edited in a menu-driven conversational mode.

Although this operation is not and should not be as recurrent as the use of the polisher itself, it is the most qualifying aspect of the Toolpack/1 polishing mechanism. Therefore, we will discuss in detail the structure and the editing of the options.

The available options are grouped into eight sections corresponding to the following Menus:

BASIC   options to activate the following actions: add sequence numbers, relabel FORMAT and statements, move FORMAT, associate CONTINUE to labels, etc.

COMMON   options giving parameters for sequence numbers, labels and indentation of statements.

UNCOMMON   options giving parameters for margins and justification for both statements and comments, as well as for formatting comments.

CONVERSION   options controlling the use of upper case and lower case characters for keywords, identifiers, strings and Format fields.

BLANK_LINES   options controlling source segmentation by means of blank lines.

LINE_BREAK   options controlling statement breaking after different tokens. The priority varies for 3 levels of parentheses.

SPACING1   options controlling insertion of spaces before each token at 3 levels of parentheses.

SPACING2   options controlling insertion of spaces after each token at 3 levels of parentheses.

When executed, the polish options editor ISTPO prompts the user for the name of the option file, then editing may start. The available commands are:

Exit, Help, Menu, Next, Query, Quit, Read, Write and "?".

Help followed by the name of a command provides information about the function performed by that command.

The command *Menu* without operand enters the menu named DIR which shows the directory of the available option sections. Each section forms a menu identified by the section name. The menus are organized as a circular set, so that the user may type the command *Next* to enter the successor of the current menu, DIR being the successor of the menu containing the last option section (namely SPACING2).

The command *Menu* followed by the name of a specific menu enters that menu directly.

The command *Query* turns on and off a safety mechanism which asks for confirmation each time one option value is changed.

The commands *Read* and *Write* allow for accessing more than one option file during the same ISTPO session.

The commands *Exit* and *Quit* terminate the session either updating the current option file or not.

Additionally there is the "change-value" command which has the following syntax:

<keyword> = <value>

where <keyword> must match one of the identifiers displayed at the terminal under one of the available menu (DIR exlcuded) and <value> must agree in type with the value to be replaced.

6.   When, How and Why to Change Polishing Options?

Potentially, new option values might be set by each user as he likes and before each run of the polisher ISTPL. However, we strongly advise people responsible for providing a Toolpack/1 user service in their own sites, to hide the option file from the end user. Those people should carefully think of a set of values which implements the presentation rules considered the most appropriate. Then, before starting the experimental use of Toolpack/1 tools, they can provide an option file with a standardized name (e.g. PL.OPT), permanently available to all users of the Portable File Store PFS. Modifications to those values may cautiously occur during the initial user testing of the polisher; a phase in which a stable image of the option file has to be reached. However, stable means neither static nor frozen. Subsequently, some option value may still be adjusted when there are proven needs or justifications for a changed parametrization of the target source presentation.

The way in which option defaults and/or previously defined values may be replaced by newly chosen ones is provided by the execution of the tool ISTPO. During a terminal session, the option editor is run conversationally. After having entered the name of the option file the user may start a dialogue he will drive according to his own specific purpose, by using the circular set of menus and the available commands.

The most articulated approach is appropriate for the non-experienced user and, in any case, during the first access to the option file. The user enters the menu DIR and then, by using the command *Next*, he goes through the full set of option sections. For any option to be modified he issues the "change-value" command, thus providing the value which will redefine the referenced option. When the modification cycle is completed for all the option sections, he terminates the editing by using the command *Exit*.

If a user already knows how options are structured in sections, a more convenient approach is to enter the selected menu by the command *Menu* followed by the specific section name and then to change as many values as necessary. Other menus may be entered in any order until exiting from ISTPO.

The quickest approach is to aim directly at the target options to be modified. This may be appropriate when a limited set of known options are going to be assigned a different value. In this case the user enters, in any order, the appropriate "change-value" commands followed by the *Exit* command.

Since the motivations concerning "when" and "why" to redefine options interact with one another, we will consider reasons and opportunities for changing option values at different moments. Redefinition of some default values (option defaults) is primarily necessary if the presentation style we wish to enforce does not match the one initially obtained by applying the option file unchanged. Redefinition of a few options during the testing period may be justified, especially in case of a constructive users feedback. On the contrary, our opinion is that, during normal service, only well documented reasons for revising the adopted presentation rules may justify modifications to the option file. One such reasons could be an inter-institutions agreement to promote the use of a commonly defined style in order to achieve a uniform – and above all a sound – formal presentation of Fortran 77 programs over a wider user community. Anyway, for whatsoever reason changes to the polish option file should have to occur in a site where ISTPL is currently used, the revised way of working should be reviewed according to a change management policy, well documented and announced in advance to both local and remote users.

7.   Conclusions

Some concluding remarks may be drawn on the efficiency of polishing source programs in the environment created by Toolpack/1 tools. Broadly speaking, this is one of the most favorable applications since only two basic tools are involved, ISTLX and ISTPL and when used as individual tools they interface by means of two files only, that is the token stream and the comment stream.

It is well known that tools based on the original version of the TIECODE library are generally slow because string-handling is mapped on integers in order to provide a fully portable solution.

As a result of a NAG investigation [25] we may summarize the following figures concerning the polishing process on a VAX 11/780 under VMS. In a VMS environment, by using the specific TIEVMS library, rather than the TIECODE library, polishing is speeded-up broadly by a factor of five.

In particular, considering the speed of the polishing process normalized to that of the host compiler, the reported performances for TIEMVS are:

0.25 for Release 1.1 of Toolpack/1;

0.49 for the fragemented tools (ISTLX + ISTPL) of Release 2.1;

0.67 for the monolith ISTLP of Release 2.1.

On one hand we may notice that the improvements introduced by Release 2 are of great satisfaction for Toolpack users, on the other hand the need for a specific implementation on every machine range is self-evident.

Fortran 77 Transformers

Malcolm J. Cohen
NAG Ltd
Oxford, U.K.

"Man is a tool-making animal."

— Benjamin Franklin,
in *J. Boswell, Life of Johnson,* 7th April 1778

Abstract. This chapter describes some of the Fortran 77 transformation tools included in the second release of Toolpack/1. The tools described are the Declaration Standardiser, Precision Transformer and Structurer. The kind of transformations performed by these tools is described in detail, and some simple examples of use are given.

1.   Introduction

This chapter describes the design features and use of three Fortran 77 transformation tools included in the second release of Toolpack/1. Two of these tools are designed to be meaning-preserving; that is, the output programs from the tool should produce the same results as the input programs when executed. These are the tools ISTDS (Declaration Standardiser) and ISTST (Structurer). The remaining tool, ISTPT, alters only the arithmetic precision of the user's program.

These tools all use the Toolpack/1 lexical and syntactic analysers described in the Chapter "Analysis Tools for Fortran 77".

Other transformation tools are available in the Toolpack/1 tool suite. Description of some of them is deferred to the next two chapters. In particular, a set of simple transformers is presented in the Chapter "Workshop on Fortran Transformations", while a special class of transformations for enhancing DO loop performances is dealt with in the Chapter "DO Loop Transforming Tools".

A. A. Pollicini (ed.), Using Toolpack Software Tools, 131–147.
© *1989 ECSC, EEC, EAEC, Brussels and Luxembourg.*

2. The Declaration Standardiser

The declaration standardiser, ISTDS, has two possible modes of operation.

mode=rebuild_declaratives

In this mode, which is the default, ISTDS rebuilds the declarations according to the template described below.

mode=declare_untyped_names

In this alternative mode ISTDS simply adds type statements for all implicitly typed names, following the comment:

```
C ..  Previously untyped names ..
```

If this comment already exists in the user's program the declarations will be added following it, otherwise the comment and new declarations will be added at the end of the existing declarations (if any).

Example DS-1 shows the effect of this mode of operation.

Example DS-1

Input program:

```
PI=3.141592653589793
PRINT *,'Input radius:'
READ *,RADIUS
AREA=PI*RADIUS**2
CIRCUM=2*PI*RADIUS
PRINT *,'Area = ',AREA,'  Circumference = ',CIRCUM
END
```

Output program:

```
C      .. Previously untyped names ..
       REAL AREA,CIRCUM,PI,RADIUS
C      ..
       PI = 3.141592653589793
       PRINT *,'Input radius:'
       READ *,RADIUS
       AREA = PI*RADIUS**2
       CIRCUM = 2*PI*RADIUS
       PRINT *,'Area = ',AREA,'  Circumference = ',CIRCUM
       END
```

2.1. The Program Template

The template used by ISTDS for the declarations of each program-unit is as follows:

(1) optional program-unit header (with preceding comments)

(2) comments before the first "declarative section"

(3) some number (possibly zero) of "declarative sections"

(4) an optional "Executable Statements" section header

(5) comments immediately preceding the first executable statement

There are no IMPLICIT or DIMENSION statements in the output from ISTDS. No IMPLICIT statements are needed because all names are explicitly typed. All arrays have their dimensions declared either in a common block declaration or in the type statement.

Each "declarative section" consists of a declarative section header of the form

```
C ..  Declarative section ..
```

followed by comments (possibly none), the declarations in that section and optionally a section trailer. ISTDS ignores case differences when detecting section headers, and converts the case to the standard form on output. A section trailer has the form

```
C ..
```

The names of the declarative sections are listed in the Appendix to this chapter, together with their section number (which may appear in some error messages produced by ISTDS). Note that statement function definitions and DATA statements are considered to be declarations and so are placed into their own declarative sections; these sections will always be the last declarative sections to be produced.

Type statements in a declarative section are produced in the order specified by the order option. The default order is: DOUBLE COMPLEX, DOUBLE PRECISION, COMPLEX, REAL, INTEGER, LOGICAL then CHARACTER. The names in each type statements are ordered alphabetically (digits are collated before letters). If both scalar and array declarations are being produced in the same section (this can only occur in the "Common blocks" section) separate type statements will be produced for each.

If there are already some section headers in the user's program, any new declarative sections added by ISTDS will be added following those already existing (except for statement function definitions and DATA statements, which are always last if they occur). The order of occurrence of existing declarative sections is preserved.

If there are no section headers then all comments occurring after the program-unit header and before the last declarative statement will appear between the program-unit header and the first declarative section. Comments between the last declaration and the first executable statement will continue to appear in that position; if the "Executable Statements" header is being inserted it will appear before these comments.

Example DS-2

Input program:

```
      PROGRAM ALPHA
C1
      CHARACTER*26 ALPHAB
      CHARACTER RESULT(26)
      DATA ALPHAB/'ABCDEFGHIJKLMNOPQRSTUVWXYZ'/
C2
      DO 100 I = 1,LEN(ALPHAB)
         K = LEN(ALPHAB) - I + 1
         RESULT(I) = ALPHAB(K:K)
  100 CONTINUE
      PRINT *,'ALPHABET = ',ALPHAB
      PRINT *,'BACKWARDS = ',(RESULT(I),I=1,26)
C3
      END
```

Output program:

```
      PROGRAM ALPHA
C1
C     .. Local Scalars ..
      INTEGER I,K
      CHARACTER ALPHAB*26
C     ..
C     .. Local Arrays ..
      CHARACTER RESULT(26)
C     ..
C     .. Intrinsic Functions ..
      INTRINSIC LEN
C     ..
C     .. Data statements ..
      DATA ALPHAB/'ABCDEFGHIJKLMNOPQRSTUVWXYZ'/
C     ..
C2
      DO 100 I = 1,LEN(ALPHAB)
         K = LEN(ALPHAB) - I + 1
         RESULT(I) = ALPHAB(K:K)
  100 CONTINUE
      PRINT *,'ALPHABET = ',ALPHAB
      PRINT *,'BACKWARDS = ', (RESULT(I),I=1,26)
C3
      END
```

2.2. Datatype Form

ISTDS has the capability of standardising the form (e.g. REAL or REAL*4) of the datatype used in type statements. This is governed by the datatype_form option, for which the following values are possible:

default

Types of differing form are considered different; e.g. REAL*4 names will continue to be declared as REAL*4 and REAL names will continue to be declared as REAL.

keywords

Datatypes which can be expressed simply as keywords will be and equivalent datatypes merged: e.g. REAL and REAL*4 names will be declared as REAL.

length_specifiers

All datatypes will include a length specification, e.g. REAL and REAL*4 names will be declared as REAL*4.

not_double_complex

This is the same as the keywords option except that double precision complex names will be declared as COMPLEX*16 instead of DOUBLE COMPLEX.

Example DS-3 shows the effects of the different values for this option.

Example DS-3

Input program:

```
REAL X
REAL*4 Y
COMPLEX*16 A
DOUBLE COMPLEX B
END
```

Output program: (1) with the default setting

```
C     .. Local Scalars ..
      DOUBLE COMPLEX B
      COMPLEX*16 A
      REAL*4 Y
      REAL X
C     ..
      END
```

Output program: (2) with **datatype_form=keywords**

```
C       .. Local Scalars ..
        DOUBLE COMPLEX A,B
        REAL X,Y
C       ..
        END
```

Output program: (3) with **datatype_form=length_specifiers**

```
C       .. Local Scalars ..
        COMPLEX*16 A,B
        REAL*4 X,Y
C       ..
        END
```

Output program: (4) with **datatype_form=not_double_complex**

```
C       .. Local Scalars ..
        COMPLEX*16 A,B
        REAL X,Y
C       ..
        END
```

2.3.   Unused Name Removal

ISTDS has the capability of removing (i.e. not producing) declarations of names which are otherwise unused. Occurrence of a name in a common block or as a dummy argument is considered to be "use" even if no other use is made.

To specify removal of unused names the option **remove_unused_names** is used. This may take the values **no** (the default), **yes** or **log**. The latter causes the names so removed to be listed on the standard error file (generally the user's terminal or batch log file).

2.4.   Other Options

The following additional options are available for controlling the output format of ISTDS.

ardicb

This option stands for ARray Dimensions In Common Blocks, and causes arrays in common blocks to be dimensioned in the COMMON statement rather than in the type statement.

chlbrk

This option causes separate CHARACTER statements to be produced for each occurring length of character name; the default is to produce a single CHARACTER statement with length specifiers after each name.

exehdr

This option causes insertion of the section header "Executable Statements" between the declarations and executable statements.

ictwcb

This option stands for Include Common Types With Common Blocks, and causes all declarations for variables in common blocks to be produced in the "Common blocks" section, following the COMMON statements, instead of in the "Scalars in Common" and "Arrays in Common" sections.

notrailers

This option prevents the production of section trailers (described above) at the end of each declarative section. Instead, the end of each declarative section is signalled by the beginning of either the next section or the first executable statement.

2.5.   Integration into Toolpack/1

The declaration standardiser, ISTDS, is integrated with five other Toolpack/1 tools. These are the include file processor (ISTIN), the lexical analyser (ISTLX), the parser (ISTYP), the polisher (ISTPL) and the include file remover (ISTUN). The two include file tools are only necessary or useful if there are "include" files in the user's program: in this case it is also necessary to specify the include_processing option to ISTDS.

The flow of data between these tools, including the include processors, is shown in Figure 1.

2.6.   Limitations

The following limitations apply to the version of ISTDS in Toolpack/1 Release 2.

- **Intrinsic Functions**

The detection of intrinsic functions is done by the parser (ISTYP) and not ISTDS. The parser only recognises the standard intrinsic functions [1] plus those described

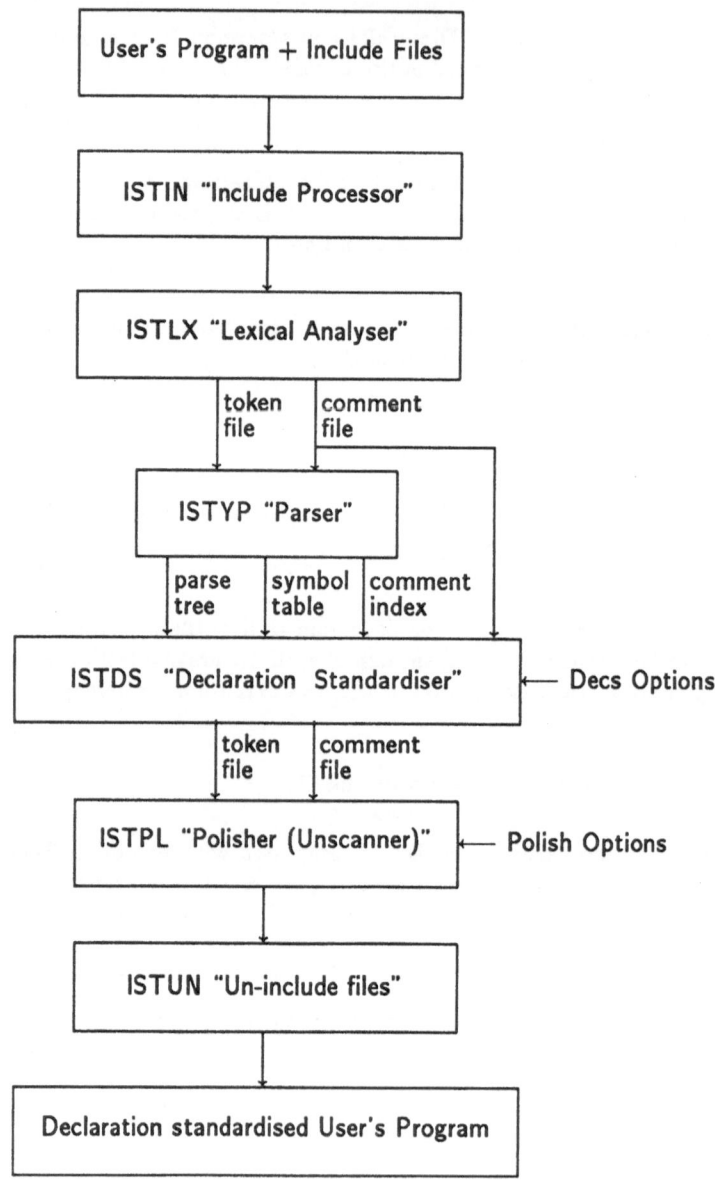

Figure 1. Declaration Standardisation: Data Flow

in Toolpack/1 Target Fortran [13]. If it is required that local intrinsic functions are recognised then the parser's intrinsic function table must be changed.

- **Adjustable Arrays**

ISTDS does not check whether variables in COMMON are used in adjustable array declarators. Since the default order of declarations puts "Array Arguments" before "Scalars in Common" illegal Fortran would be generated if the implicit type of the common variable was not INTEGER.

This can be avoided by manually inserting the "Scalars in Common" section header before processing the program with ISTDS.

- **Continuation Lines**

ISTDS cannot check for the maximum number of continuation lines being exceeded because it only uses the parse tree and token stream representations of the user's program. Thus the code it generates will be illegal if either

(1) there are too many items of the same datatype and section to fit into a single type statement, or

(2) a common block has too many items to fit into one COMMON statement.

In these cases an error will be reported by the polisher (ISTPL) when converting the program back to source text and the user must manually edit the output to split up the long statements.

3. The Precision Transformer

The Toolpack/1 precision transformer, ISTPT, transforms the arithmetic precision of a user's program from single-precision to double-precision or vice versa. In particular, the transformation occurs between REAL and DOUBLE PRECISION, and between COMPLEX and DOUBLE COMPLEX.

As ISTPT operates on the parse tree, which is an abstract syntax tree (and not a concrete syntax tree), some optional syntax may be introduced into or deleted from the user's program. These are the optional commas in CHARACTER and DO statements (which are not produced), the optional commas in FORMAT statements (these are produced), the control-information specifier 'FMT=' in I/O statements (this is produced) and the optional empty parentheses in a SUBROUTINE statement for a subroutine subprogram with no arguments (not produced).

3.1. Details of Precision Transformation

The details of the transformations performed by ISTPT are listed below. In addition to these transformations, ISTPT checks all EQUIVALENCE statements in the user's program; if the meaning of these changes under transformation an error message is produced.

(1) The keyword REAL is changed to/from DOUBLE PRECISION, and the keyword COMPLEX is changed to/from either DOUBLE COMPLEX or COMPLEX∗16 (according to the setting of the option dcform).

(2) Declarations are added for implicitly typed names which have changed precision.

(3) Real constants are changed to/from double precision constants, adding or deleting 'D0' from the end of decimal constants where appropriate.

(4) Single-precision complex constants are transformed to/from double-precision complex constants. If both parts of a complex constant are integer, the suffix 'D0' is added to the constant for the real part.

(5) The E-format edit descriptor is changed to/from the D-format edit descriptor, except for the form 'Ew.dEe' which has no D-format counterpart.

(6) Single-precision non-generic intrinsic function names replace/are replaced by double-precision intrinsic function names, except for DPROD, MAX1, AMAX0, MIN1, AMIN0 and AIMAG.

(7) DPROD is replaced by multiplication (with any necessary addition of parentheses). Occurrence of DPROD in a double-precision source program is considered an error.

(8) MAX1(...) is replaced by INT(MAX(...)) and MIN1(...) is replaced by INT(MIN(...)), when converting to double-precision.

(9) AMAX0(...) is replaced by DBLE(MAX(...)) when converting to double-precision. DBLE(AMAX0(...)) is replaced by AMAX0(...) when converting to single-precision. AMIN0 is handled similarly.

(10) AIMAG(...) is transformed to DBLE(AIMAG(...)) when converting to double-precision. DBLE(AIMAG(...)) is transformed to AIMAG(...) when converting to single-precision.

Example PT-1

Converting to Double-precision

Input program:

```
      REAL PI,RADIUS,AREA,CIRCUM
      PI = 3.141592653589793
      PRINT 9000
      READ *,RADIUS
      AREA = PI*RADIUS**2
      CIRCUM = 2.*PI*RADIUS
      PRINT 9100,CIRCUM,AREA,AMAX1(AREA,CIRCUM)
 9000 FORMAT (' Input Radius:')
 9100 FORMAT (' Circumference = ',F10.6,'  Area = ',E12.6,'  Max = ',
     +        E12.6)
      END
```

Output program:

```
DOUBLE PRECISION PI,RADIUS,AREA,CIRCUM
PI = 3.141592653589793D0
PRINT 9000
READ *,RADIUS
AREA = PI*RADIUS**2
CIRCUM = 2.D0*PI*RADIUS
PRINT 9100,CIRCUM,AREA,DMAX1(AREA,CIRCUM)
9000 FORMAT (' Input Radius:')
9100 FORMAT (' Circumference = ',F10.6,'  Area = ',D12.6,'  Max = ',
   +      D12.6)
END
```

3.2.   Limitations

ISTPT recognises the same set of intrinsic functions as the parser, ISTYP. If it is desired to add site-specific intrinsic functions and their transformations between precision these transformations must be added to ISTPT and the new functions added to the list recognised by ISTYP.

ISTPT is not always able to remove declarations of intrinsic functions which are no longer used. In this case an informational message is produced on the standard error file.

If the name of an intrinsic function which is inserted by one of the transformations has already been used for some other purpose (e.g. as a local variable), ISTPT will abort with the error "Inconsistent symbol types". If it has been used as the name of a user-written external function then ISTPT will simply produce incorrect code.

3.3.   Integration into Toolpack/1

The precision transformer, ISTPT, is integrated with three other Toolpack/1 tools. These are the lexical analyser (ISTLX), the parser (ISTYP) and the polisher (ISTPL).

The flow of data between these tools is shown in Figure 2.

4.   The Structurer

ISTST is the program structuring tool in Toolpack/1 Release 2. It rebuilds the flow of control within each program-unit processed to a standardised form. This is done

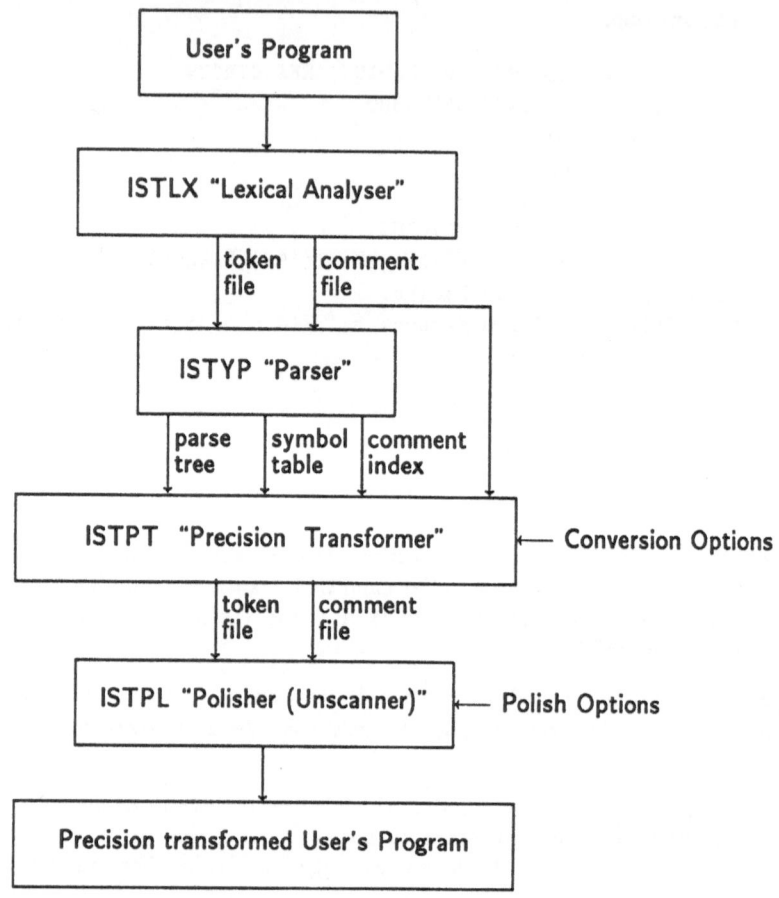

Figure 2. Precision Transformation: Data Flow

by analysing the parse tree of the program, building a flowgraph (eliminating explicit transfers of control), and producing an equivalent program, re-introducing explicit transfers of control according to a prescribed scheme.

ISTST, like the Unix program *struct* , is based on the structuring algorithm described in [3]. Unlike *struct,* however, its output and input languages are both Fortran 77.

4.1.   Recognising Structure in a Flowgraph

The basic unit of structure in the flowgraph is the loop. A flowgraph is called *reducible* if each loop is entered in only one place; an irreducible flowgraph requires a jump

(GOTO) into the middle of a loop. Irreducible flowgraphs are detected by ISTST and output unchanged. An irreducible flowgraph cannot be made reducible without copying code or introducing extra control variables (which we have decided not to do).

4.2.   The Structuring Operation

The output program from ISTST obeys the following structures:

(1) There is a path from each statement within the extent of a loop back to the head of the loop.

(2) A statement should appear within the THEN-clause of an IF-THEN-ELSE if it can only be reached from that clause and it is nested within the innermost loop containing the IF. Similar conditions hold for statements within an ELSE clause.

(3) If the statements within an IF block can only reach a single point outside the block, and this point does not appear immediately after the IF (e.g. if it is the head of the innermost containing loop) then there should be a GOTO following the end of the IF-block rather than individual GOTO's within each clause.

(4) The head of a loop appears in the output program either as a DO statement (if it is a DO-loop) or as a CONTINUE statement. A DO-loop is terminated with a CONTINUE statement.

4.3.   Handling of Unusual Constructs

- **Arithmetic IF statements**

Arithmetic IF statements are altered by the preprocessing pass into logical IF statements unless the three transfer labels are different: in this case the arithmetic IF is left unchanged and will appear in the same form in the output.

- **Assigned GOTO statements**

All assigned GOTO statements are converted into computed GOTO statements, and the ASSIGN statements changed to assignment statements. The *fall-through* option of the computed GOTO is used for the first alternative (ASSIGN) seen.

If there is only one possible target for the assigned GOTO, it is converted into a simple GOTO (which might be removed during structuring). If there are no possible targets for the assigned GOTO, ISTST terminates with a fatal error message.

ASSIGN statements which reference FORMAT statements (i.e. for use in READ and WRITE) are left unchanged.

- **RETURN statements**

Simple RETURN statements are treated by ISTST as transfers of control to the END statement. As such they may disappear or be merged in the output. Similarly, any transfer of control to the END statement (of a SUBROUTINE or FUNCTION) in the output is represented by a RETURN (not a GOTO).

Alternative RETURN statements are treated as ordinary straight-line code followed by an implicit transfer of control to the END statement.

- **STOP statements**

STOP statements are treated by ISTST as ordinary straight-line code; i.e. they are not "understood", and code following them will continue to follow them. This may on occasion lead to STOP statements being followed by a GOTO statement.

- **Dead code**

Executable statements (except for END, END IF, etc.) which follow an unconditional transfer of control are not reachable and are thus *dead*. Since there will be no path to them in the flowgraph ISTST will automatically remove them. A warning message is produced for each such unreachable statement informing the user that they have been removed.

- **Comments**

Comments which are either before or embedded within an executable statement will remain with that statement in the output. Comments associated with a flow-of-control statement (e.g. CONTINUE, GOTO, or ENDIF), are treated as separate executable statements for the purpose of flowgraphing. (This is because flow of control statements are eliminated during flowgraphing and only added to the output where necessary).

Example ST-1

Input program:

```
        SUBROUTINE XXX
        A=1
        IF (A.EQ.B .OR. C.EQ.D) GOTO 200
        DO 10 C=1,10
            D=1
            IF (E) GOTO 220
   10   CONTINUE
  200   DO 210 F=1,10
            G=1
  210   CONTINUE
  220   CONTINUE
        END
```

Output program:

```
        SUBROUTINE XXX
        A = 1
        IF (A.NE.B .AND. C.NE.D) THEN
            DO 10 C = 1,10
                D = 1
                IF (E) RETURN
   10       CONTINUE
        END IF
        DO 20 F = 1,10
            G = 1
   20 CONTINUE
        END
```

Example ST-2

Input program:

```
        READ *,A
        IF (A) 601,601,602
   601 PRINT *,'A.LE.0'
        IF (A) 701,702,701
   602 PRINT *,'A.GT.0'
   701 PRINT *,'A.NE.0'
   702 PRINT *,'A IS ',A
        END
```

Output program:

```
        READ *,A
        IF (A.GT.0) THEN
            PRINT *,'A.GT.0'
        ELSE
            PRINT *,'A.LE.0'
            IF (A.EQ.0) GO TO 10
        END IF
        PRINT *,'A.NE.0'
   10 PRINT *,'A IS ',A
        END
```

4.4.   Integration into Toolpack/1

The structurer, ISTST, is integrated with two other Toolpack/1 tools. These are the lexical analyser (ISTLX) and the parser (ISTYP). The polisher (ISTPL) is included in the structurer execution (monolithic integration).

The flow of data between these tools is shown in Figure 3.

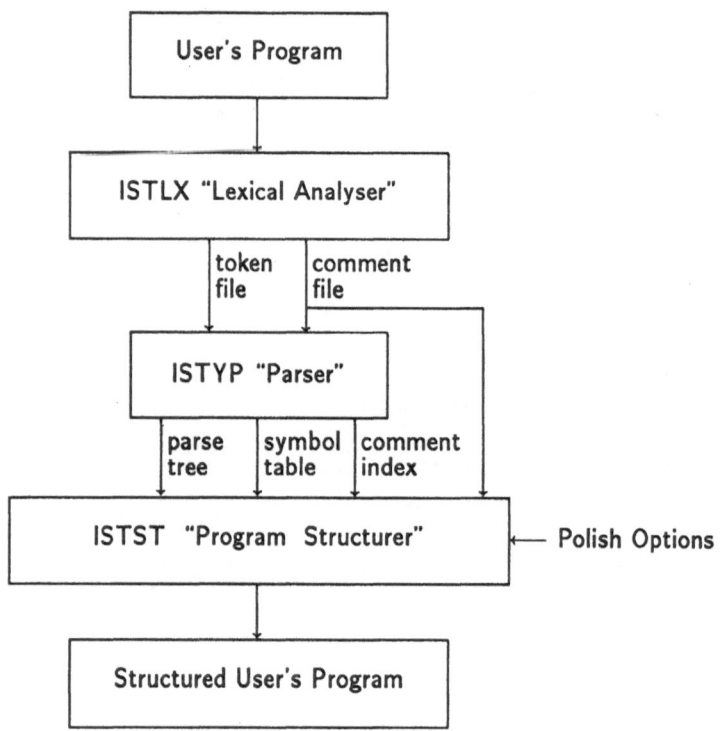

Figure 3. Program Structurer: Data Flow

Appendix:
ISTDS Section Names

1 Parameters
2 Scalar Arguments
3 Array Arguments
4 Function Arguments
5 Subroutine Arguments
6 Scalars in Common
7 Arrays in Common
8 Local Scalars
9 Local Arrays
10 External Functions
11 External Subroutines
12 Intrinsic Functions
13 Common blocks
14 Entry Points
15 Statement Functions
16 Equivalences
17 Save statement
18 Statement Function Definitions
19 Executable Statements
20 Previously untyped names
21 DATA Statements

Workshop on Fortran Transformations

Malcolm J. Cohen
NAG Ltd
Oxford, U.K.

"Man is a tool-using animal ...
Without tools he is nothing, with tools he is all."

— Thomas Carlyle,
Sartor Resartus, bk.I, ch.5, 1838

Abstract. This chapter describes the structuring tool ISTST in further detail, as well as several of the smaller Fortran 77 transformation tools from Toolpack/1 Release 2. These smaller tools are oriented towards performing simple regular transformations on user programs which would be prone to error if done manually.

1.   Introduction

The first part of this chapter describes the internal operation of the structuring tool ISTST. This tool was introduced in the Chapter "Fortran 77 Transformers". The rest of this chapter describes several of the smaller Fortran 77 transformation tools included in the second release of Toolpack/1.

The smaller tools are:

(1) a tool for making references to specific intrinsic functions into references to generic intrinsics (ISTGI),

(2) a tool for rearranging expressions so that compilers need less stack space to compile them (ISTME),

(3) a tool for joining and splitting strings in FORMAT statements (ISTJS), and

(4) a tool for ensuring PARAMETER consistency throughout a whole program.

149

A. A. Pollicini (ed.), *Using Toolpack Software Tools*, 149–160.
© 1989 ECSC, EEC, EAEC, Brussels and Luxembourg.

2.   Internal Operation of ISTST

The processing by ISTST of each program-unit is split into the following three passes.

2.1.   First Pass – Parse Tree Canonicalisation

The first pass of ISTST is performed by the library routine ZFCAPU (from the ACCESS supplementary library). This routine performs the following conversions:

(1) Arithmetic IF statements are converted into logical IF statements (possibly with a following GOTO) except in the case where all three labels are different.

(2) All DO loops are made to end on unique CONTINUE statements.

(3) COMMENT statements are added to hold comment information immediately prior to statements which shall be removed during flowgraphing: that is, CONTINUE, unconditional GOTO, ENDIF and ELSE. (This is only done where there are actually some comments).

(4) The extended data (see ZYSTXF and ZYGTXF) for each statement-level node in the parse tree is set to the original statement number for that statement.

2.2.   Second Pass – Flowgraph Creation

- **Flowgraphing Principles**

A flowgraph is a representation of the executable part of a program. It is a directed graph with each node representing a single executable statement and arcs representing transfers of control. The arcs from a particular node are its *outarcs* and the arcs going into a particular node are its *inarcs*.

There are three types of statement which are not represented by flowgraph nodes: declarative statements, (unconditional) control-flow statements and null statements. These control-flow statements are the unconditional GOTO, ELSE and RETURN (without an alternate specifier). The null statements are CONTINUE and ENDIF.

There are four types of flowgraph node. These are:

(1) The straight-line code node. This node has exactly one outarc. It represents any executable statement (except for END) which does not branch.

(2) The exit node. This node has no outarcs. It represents the END statement.

(3) The if node. This node has exactly two outarcs. It represents either an IF-conditional (except arithmetic IF) or a DO statement.

(4) The case node. This node has at least two outarcs. It represents an arithmetic IF, computed GOTO, I/O statement with 'ERR=' and/or 'END=' specifiers, a subroutine call with alternate return specifiers or a dummy node inserted at the

beginning of the flowgraph of a subprogram which has ENTRY points: this dummy node is distinguished by having a parse tree node pointer of "-2".

- **Flowgraph Equivalence**

For the purposes of structuring, we may consider two programs to be equivalent if they have the same flowgraphs. This definition of equivalence precludes the use of the techniques of code copying and introduction of control variables by the structurer.

- **Flowgraph Creation Phases**

The process of creating the flowgraph contains several phases. The first phase creates the raw flowgraph in the form described above.

During this subphase any assigned GOTO statements are converted into computed GOTO statements, by scanning the parse tree, converting each ASSIGN into an assignment and associating the non-negative integers with the labels encountered in the appropriate ASSIGN statements. The *fall-through* exit from the computed GOTO thus becomes the first of the labels encountered.

The second phase constructs the *spanning tree* using a depth-first search. This numbers the nodes so that nodes which are deeper in the tree have higher numbers. If the flowgraph is *reducible*, this numbering scheme will uniquely identify forward and backward arcs in the flowgraph: the backward arcs will be those where the *number* of the node (according to the depth-first search) decreases.

The third phase inserts extra nodes into the flowgraph for every loop. Each node which is the target of a backward arc is the head of a loop; the new node, called a *repeat* node, is inserted immediately prior to the existing loop header(s) and all backward arcs modified to point to this new node instead. A *repeat* node can be distinguished from other flowgraph nodes by its parse tree index of "-1" (all valid parse tree node pointers being positive).

The fourth phase walks the flowgraph setting each node's *head* pointer to the head of the innermost loop containing that node. At this point an irreducible flowgraph is detected and no further processing performed.

The fifth phase counts the number of forward arcs entering each node.

The sixth phase determines each node's dominator: this is the closest node through which control must pass to reach that node.

The seventh and final phase uses the information gathered in the previous phases to determine the *follow* sets of each node. A node is said to *follow* another if it is dominated by that node (or a node nested with it if it is a repeat) and is at the same level of nesting (i.e. it is not outside an inner loop or nested within that node).

2.3.   Third Pass – Structured Output

The final pass of ISTST produces the structured output from the information stored in the flowgraph. The basic structure of the algorithm for producing the structured code can be seen in the following (recursive) pseudo-code subroutine "STRUCT", which is called on the initial node of the flowgraph). For simplicity we shall ignore the existence of *case* nodes (i.e. computed GOTO etc.): they are processed similarly to IF nodes.

```
SUBROUTINE STRUCT(X)

IF (X is straight-line code) THEN
    Output the statement
ELSE IF (X is a repeat node) THEN
    Output a labelled CONTINUE statement
    IF (the successor to X is not in any FOLLOW set) THEN
        CALL STRUCT(the successor to X)
    END IF
ELSE IF (X is an IF node) THEN
    Output an IF-THEN statement
    IF (the true arc is forwards and not in any FOLLOW set) THEN
        CALL STRUCT(the 'true' arc)
    END IF
    Output an ELSE statement
    IF (the false arc is forwards and not in any FOLLOW set) THEN
        CALL STRUCT(the 'false' arc)
    END IF
    Output an ENDIF statement
END IF
IF (the next node which will be output is not the one to which
    control should pass next) Output a GOTO statement
For each node Y in the FOLLOW set of X, CALL STRUCT(Y)

END
```

Most of the improvements to the basic algorithm are limited to IF statements since Fortran does not provide any looping constructs apart from DO-loops and GOTO statements. The main improvements are listed below.

• **Elimination of the ELSE clause**

If the ELSE clause of the IF is going to be empty then it is not output. If the THEN clause is empty then the IF-condition is negated so that the ELSE clause can be eliminated.

If the ELSE clause has been eliminated and there is only one statement in the THEN clause, a logical IF statement is produced instead of a block IF.

- **Elimination of internal GOTOs**

If there is only one node outside the IF-block which can be reached from within the IF-block, and that node is not the next to be output, then a GOTO is output following the ENDIF. This prevents the following from being produced:

```
100    CONTINUE
       IF (A) THEN
           IF (B) THEN
               ...
               GOTO 100
           ELSE
               ...
               GOTO 100
           END IF
       END IF
       ...
       END
```

Instead, a single "GOTO 100" is produced following the block beginning with the statement 'IF (B) THEN'.

- **Production of ELSE-IF**

If one clause of the IF statement consists entirely of another IF-node and those nodes which are nested within it, then an ELSE IF is generated instead. I.e., the following constructs are detected and transformed as shown:

```
IF (A) THEN                            IF (A) THEN
    ...1                                   ...1
ELSE                                   ELSE IF (B) THEN
    IF (B) THEN        ===>                ...2
        ...2                           END IF
    END IF
END IF
```

```
IF (A) THEN                            IF (.NOT.A) THEN
    IF (B) THEN                            ...2
        ...1                           ELSE IF (B) THEN
    END IF            ===>                 ...1
ELSE                                   END IF
    ...2
END IF
```

3.   ISTGI – Make Intrinsics Generic

ISTGI operates on the symbol table produced by the Toolpack/1 parser ISTYP (described in the Chapter "Analysis Tools for Fortran 77"), and produces a new symbol table with specific intrinsic function names replaced by generic function names. This method of conversion is very fast and simple and will convert most uses of non-generic intrinsic functions. To produce new source code, the Toolpack/1 parse tree flattener ISTYF [10] is used to create a token stream that can be used as input to the polisher ISTPL.

3.1.   Conversion Restrictions

There are three instances where ISTGI cannot convert a function to its generic form:

(1) when a function has been used as an actual parameter to another function or subroutine,

(2) when the function has been explicitly typed, or

(3) when the generic function name it would have been converted to has been used for some other purpose within the program unit, for example as an array name.

In these cases ISTGI will issue a warning message saying that it cannot convert the function for that reason.

3.2.   Symbol Table Attributes

ISTGI will preserve all symbol table attribute information except when a duplicate symbol is produced. This can occur either because two different specific intrinsic functions are mapped onto the same generic function or because the generic function name was used as well as one of the specific function names. In this case ISTGI will produce a warning message, indicating that the parse tree and symbol table will need to be flattened (using ISTYF) and reparsed (using ISTYP) to recover the attribute information.

Note that when reading the symbol table, any errors previously reported by ISTYP will be reported again on the standard error channel. These errors are not, however, copied to the new symbol table.

3.3.   Integration into Toolpack/1

The diagram of Figure 1 shows the integration of ISTGI with the tools ISTLX, ISTYP, ISTYF and ISTPL.

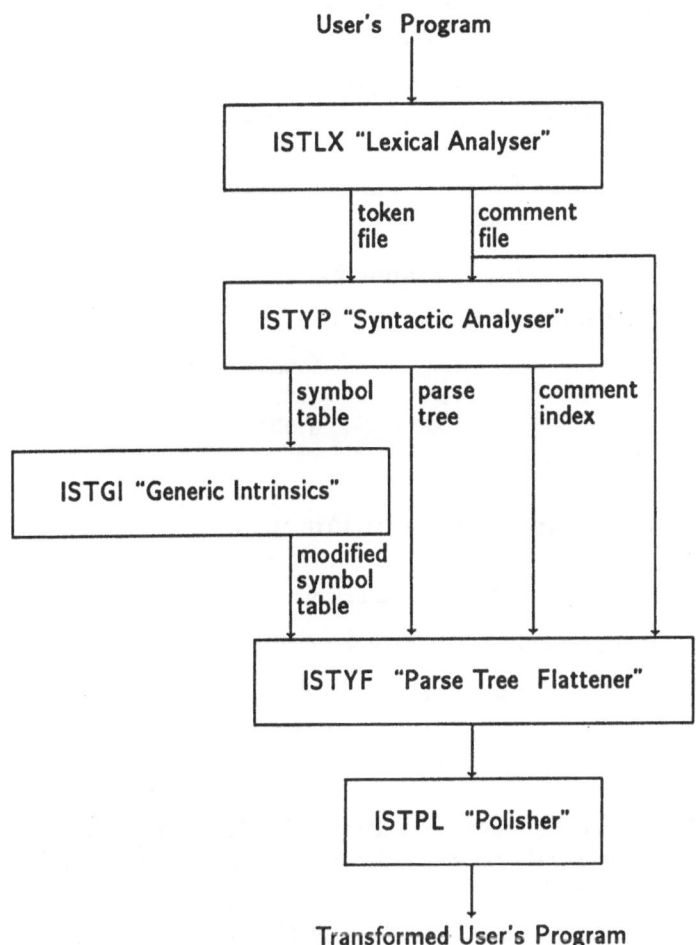

Figure 1. Integration of ISTGI into Toolpack/1

4.   ISTFR – Real Constant Modifier

This tool converts REAL, DOUBLE PRECISION, COMPLEX and DOUBLE COMPLEX constants to a consistent form. Input to the tool is a token stream from the lexical analyser, ISTLX; output is another token stream which may be converted to source form with the polisher, ISTPL.

The following actions are performed for each constant:

(1) The exponent character (if present) is converted to upper case.

(2) Any leading plus sign on the exponent is removed.

(3) The exponent is removed if it is equal to zero (unless it is a double-precision constant).

(4) Ensure that the mantissa contains a decimal point with at least one digit before and after it.

Example FR-1

Constants Converted   From: To:

1e3   1.0E3
.25   0.25
.0d0   0.0D0

5.   ISTJS – Join Strings in Formats

This tool modifies strings in FORMAT statements.

The following may be, or are, done in the order specified:

(1) Convert holleriths to string constants (optional).

(2) Convert all X edit descriptors to string constants (optional).

(3) Convert X edit descriptors adjacent to existing strings into string constants (optional).

(4) Convert blanks at the beginning or end of strings that are preceded or followed by X edit descriptors into the associated X edit descriptor, e.g. (2X,' PASS') becomes (3X,'PASS'), (optional).

(5) Combine consecutive string constants into single values (compulsory).

(6) Combine consecutive X edit descriptors (compulsory).

(7) Delete optional commas preceding or following '/' edit descriptors (compulsory).

The operation of the tool is controlled by the option string as follows:

'-' No-effect

'a' Toggle conversion of X edit descriptors to strings. Unlike the 'x' option this only converts X edit descriptors to strings of blanks if they are adjacent to an existing string.

'h' Toggle conversion of holleriths, default TRUE.

'j' Toggle combining of X edit descriptors and leading or trailing blanks in strings, default TRUE.

'x' Toggle conversion of X edit descriptors, default FALSE.

Option characters may be combined in any quantity and in any order in the option string. Each legal character detected toggles the appropriate flag.

NOTE: if both X and J flags are true then the effect is the same as X=true and J=false.

5.1.   Operation

The input to ISTJS is the user's program in token stream form (as produced by the lexical analyser, ISTLX) and an option string. The output from ISTJS is the modified user program, again in token stream form.

Example JS-1

Input program:

```
      PRINT 9000
 9000 FORMAT(1X,' HELLO',' THERE')
      END
```

Output program:

```
      PRINT 9000
 9000 FORMAT (2X,'HELLO THERE')
      END
```

6.   ISTME – Manipulate Expressions

This tool modifies expressions so as to minimise the depth of stack required for compilation. With commutative operators (i.e. addition and multiplication) the subexpression of least depth is placed first. Other operators are left unchanged, as are parentheses.

The input to ISTME is the parse tree produced by the Toolpack/1 parser ISTYP (none of the other files - the symbol table, comment index or comment file - are used). The output from ISTME is another parse tree. This may be used to replace the original parse tree in the input to any other parse tree-based tool; to turn the output back to source text the flattener (ISTYF) and polisher (ISTPL) are used.

Example ME-1

Input program:

```
A = (B+(C+(D+(E+1)))) + 2
A = 2 + ((((E+1)+D)+C)+B)
END
```

Output program:

```
A = ((((E+1)+D)+C)+B) + 2
A = ((((E+1)+D)+C)+B) + 2
END
```

7.   ISTPP – Program Parameter Verifier

This tool ensures that a set of common PARAMETERs have the same values throughout a Fortran 77 file. This provides an alternative form of processing to the use of macro preprocessing. PARAMETER values are defined in source embedded directives (described in [13]). These may be placed in the code to be processed or in a library file. The source embedded directives are of the following form:

    *PP <name> = <value>

ISTPP then searches the input file for PARAMETERs called <name> and ensures that they are set to <value>. If a PARAMETER is not correctly set then the PARAMETER statement is modified.

7.1.   Operation

The input to ISTPP consists of the user program in token stream form together with a number (possibly zero) of library files. The output from ISTPP is the (possibly modified) user program, again in token stream form.

All library files are read before processing the user program. The library files may have INCLUDE statements. These function similarly to those detected by ISTIN (described in the chapter "The Toolpack/1 Tool Suite") using the 'p' option. If a library file name is not supplied as part of the command-line arguments the user will be prompted repeatedly for library files until a blank line is entered. (A hyphen may be used as the library file argument to indicate that there are no library files).

Example PP-1

Input program:

```
*$pp$ FRED=3
      PROGRAM PP1
      PARAMETER (FRED=0)

      CALL PP1A(FRED)
      END
      SUBROUTINE PP1A(X)
      PARAMETER (FRED=9.9)

      PRINT *,FRED + X
      END
```

Output program:

```
*$pp$ FRED=3
      PROGRAM PP1
      PARAMETER (FRED=3)

      CALL PP1A(FRED)
      END
      SUBROUTINE PP1A(X)
      PARAMETER (FRED=3)

      PRINT *,FRED + X
      END
```

7.2.   Supplied Library Files

There are three ISTPP library files supplied in Toolpack/1 Release 2 for the use of Toolpack-compatible software tools written in Fortran 77.

ZPSLIB   contains parameter settings for various system-defined constants. These are:

    (1) status codes (e.g. OK, EOF, ERROR, ...)

    (2) file access codes (e.g. READ, WRITE, READWR)

    (3) pre-connected unit numbers (STDIN, STDOUT, STDERR, STDLST)

    (4) buffer sizes (e.g. MAXNAM, MAXLIN, MAXBUF, ...)

ZPILIB   which may be found in the GENERAL tools file. It contains parameter settings for the characters in the portable (IST) character set.

ZPTLIB   which is part of the ACCESS supplementary library. It contains the
         parameter settings for token type names (listed in the first Appendix of
         the Chapter "Analysis Tools for Fortran 77").

7.3.   ISTPP-conforming tools in Toolpack/1

The following tools in Toolpack/1 Release 2 are in standard Fortran 77 with ISTPP-
format PARAMETER's:

(1) ISTPP itself

(2) ISTP2 (a monolithic tool combining ISTLX, ISTPP and ISTPL)

(3) ISTET (tab expander)

DO Loop Transforming Tools *

Wayne R. Cowell
Mathematics and Computer Science Division
Argonne National Laboratory
Argonne, Illinois, U.S.A.

" ... sempre una mutazione lascia l'addentellato per la edificazione dell'altra."

— Niccolò Machiavelli,
Il Principe, ch.II, 1513

"A tool is but the extension of a man's hand, and a machine is but a complex tool."

— Henry Ward Beecher,
Proverbs from Plymouth Pulpit

Abstract. This chapter is an overview of three transformation tools recently contributed to the Toolpack/1 tool collection. The first tool unrolls outer DO loops, the second condenses sequences of inner DOs in unrolled outer DOs into single DO loops together with assignment statements, and the third combines sequences of assignment statements resulting from the action of the first two tools. Acting in concert with one another and with other Toolpack/1 tools, these tools transform certain classes of Fortran DO nests that typically occur in linear algebra software. The effect is to improve the performance of such software on vector machines by reducing the number of words transferred between memory and the vector registers. Emphasised in this chapter are features of the Toolpack tool-writing environment that are employed in writing such transformation tools.

* Work supported in part by the Applied Mathematical Sciences subprogramme of the Office of Energy Research, U. S. Department of Energy, under contract W-31-109-Eng-38.

161

A. A. Pollicini (ed.), Using Toolpack Software Tools, 161–180.
© 1989 ECSC, EEC, EAEC, Brussels and Luxembourg.

1.   Introduction

This chapter is a report on the design, implementation, and operation of three Fortran transformation tools that have recently been contributed to the Toolpack/1 tool collection. We shall summarise the purpose and use of these tools, but our primary motivation is to provide a case study of techniques for writing large transformation tools, in order to exhibit features of the Toolpack/1 tool-writing environment. The use of these tools is covered in more detail in [15]. An analysis of the transformations that they perform and a report of their application to a widely used collection of linear algebra software may be found in [19]. We shall assume that the reader is already familiar with Toolpack/1 transformation tools and with the Toolpack/1 tool-writing environment. Introductions to both topics can be found in the Chapters "Toolpack/1 Fortran Transformers" and "Tool Writing" respectively.

The transformation tools we study are intended for use in a programming environment for vector machines. Their purpose is to transform nests of Fortran DO loops that match certain patterns commonly found in basic linear algebra software so that the number of vector loads and stores is reduced when the transformed Fortran is compiled by a vectorising compiler. Memory access is a major bottleneck in advanced scientific computation, so that reducing its frequency results in a substantial improvement in performance.

2.   The Transformations and Their Effect

A simple example will illustrate the transformations and their effect. Suppose a machine with vector instructions has vector registers that are 64 words long. Following is a nest of DO loops that calculates the product of a $64 \times N$ matrix A and a $N \times 1$ vector B, storing the result in a 64×1 vector C which is assumed to have been initialised to 0.

```
      DO 10 J = 1,N
         DO 20 I = 1,64
            C(I) = C(I) + A(I,J)*B(J)
20          CONTINUE
10    CONTINUE
```

For convenience, assume that N is even. Then the following code is arithmetically equivalent to the above. The outer DO has been *unrolled* to *depth* 2; that is, its range has been replicated and the parameters of the DO suitably modified. As a consequence, each pass does twice as much computation but there are only half as many passes.

```
      DO 10 J = 1,N-1,2
         DO 20 I = 1,64
            C(I) = C(I) + A(I,J)*B(J)
20          CONTINUE
         DO 30 I = 1,64
            C(I) = C(I) + A(I,J+1)*B(J+1)
30          CONTINUE
10       CONTINUE
```

For any allowable value of J, say J = j, the pair of inner loops is equivalent to the following sequence of assignment statements:

```
      C(1) = C(1) + A(1,j)*B(j)
      C(2) = C(2) + A(2,j)*B(j)
                .
                .
                .
      C(64) = C(64) + A(64,j)*B(j)
C ***********************************************
      C(1) = C(1) + A(1,j+1)*B(j+1)
      C(2) = C(2) + A(2,j+1)*B(j+1)
                .
                .
                .
      C(64) = C(64) + A(64,j+1)*B(j+1)
```

where the comment of repeated "*" separates the statements into two groups from the two original inner DO loops. Observe that none of $C(2)$, $C(3)$, ..., $C(64)$ occurs in the first statement in the second group. Furthermore, $C(1)$ does not occur in the second, third, ..., last statement of the first group. Because of this independence, no quantity in the computation is changed if we move the first statement of the second group upward so that it immediately follows the first statement of the first group, giving

```
      C(1) = C(1) + A(1,j)*B(j)
      C(1) = C(1) + A(1,j+1)*B(j+1)
      C(2) = C(2) + A(2,j)*B(j)
                .
                .
                .
      C(64) = C(64) + A(64,j)*B(j)
C ***********************************************
      C(2) = C(2) + A(2,j+1)*B(j+1)
                .
                .
                .
      C(64) = C(64) + A(64,j+1)*B(j+1)
```

By a similar argument the further statements in the second group may be moved upward to give

```
C(1) = C(1) + A(1,j)*B(j)
C(1) = C(1) + A(1,j+1)*B(j+1)
C(2) = C(2) + A(2,j)*B(j)
C(2) = C(2) + A(2,j+1)*B(j+1)
             .
             .
C(64) = C(64) + A(64,j)*B(j)
C(64) = C(64) + A(64,j+1)*B(j+1)
```

which is equivalent to writing the two loops as one. Hence the nest becomes

```
      DO 10 J = 1,N-1,2
         DO 20 I = 1,64
            C(I) = C(I) + A(I,J)*B(J)
            C(I) = C(I) + A(I,J+1)*B(J+1)
20          CONTINUE
10       CONTINUE
```

We say that the sequence of DOs has been *condensed* into a single DO.

It is clear by examining the sequence of assignment statements equivalent to the condensed inner DO that the pair of assignments to any array element can be combined by substituting the first assignment of each pair into the right side of the second and eliminating the first. This is called *substitution/elimination* and results in the following transformed DO nest:

```
      DO 10 J = 1,N-1,2
         DO 20 I = 1,64
            C(I) = C(I) + A(I,J)*B(J) + A(I,J+1)*B(J+1)
20          CONTINUE
10       CONTINUE
```

An examination of the code produced by a typical vectorising compiler reveals that the original nest performs 2 vector loads and a vector store for each traversal of the inner loop, while the transformed nest performs 3 vector loads and a vector store for each traversal. Sixty-four words are moved between vector registers and memory for each load or store. The original loop is traversed N times, so $192N$ words are moved, while the transformed loop is traversed only $N/2$ times, so $128N$ words are moved.

In principle, we may further reduce the transfers between memory and vector registers by unrolling to greater depth. Assume that the unrolling depth divides N. Then, in the example above, $96N$ words would be moved if the unrolling depth were 4, and $66N$ words if the unrolling depth were 64, the maximum meaningful depth. The actual decrease in execution time depends on the architecture of the machine – the number of vector registers, whether there are concurrent paths from memory, etc.

The assumption that the unrolling depth divides N is not essential. In general, there will be values of the DO variable not assumed by the new set of parameters, and it will be necessary to have an additional "clean-up" loop. The effect of this additional computation is usually small, especially when the limit of the DO is large.

Transformations such as these were reported in [21], [22] and [23], stimulating our interest in developing tools to perform the transformations. Data obtained from timing experiments on a CRAY-1S (see [19]) using the basic linear algebra kernels described in [21] show that tool-transformed and human-transformed kernels were about equally efficient. In both cases, the decrease in execution time for user-callable programs that called the kernels was typically 30% to 50%.

The advantage of automation is, of course, that computers are faster and more accurate than humans at performing complex but well-defined tasks. If tools to perform the above transformations are available in the programming environment for a vector computer, library routines can be written and maintained in their original, more compact, form with the task of generating transformed versions given to the tools. Programmers thus can easily generate transformed versions and perform experiments that determine the optimum unrolling depth for a particular routine on a particular machine configuration.

3.   Tool Integration in Toolpack/1

As is pointed out in the Chapter "Toolpack/1 Fortran Transformers" and in "*Toolpack/1 Introductory Guide*"[17], the Toolpack/1 lexer (ISTLX) maps Fortran text into a token/comment stream and the parser (ISTYP) maps a token/comment stream into a parse tree/ symbol table. The formatter tool (ISTPL), acting as an "unscanner", maps a token/comment stream into Fortran text, while various Toolpack/1 tools, including the declaration standardiser and the precision transformer, map a parse tree/symbol table into the token/comment stream of the transformed program. These latter are called parse tree "flattening" tools. To describe the Toolpack/1 tool collection as *integrated* is a way of characterising the interdependence among tools, as exhibited in Fig. 1.

Three tools, ISTUD, ISTCD, and ISTSB, effect the transformation of Fortran DO nests. They perform unrolling, condensing, and substitution/elimination, respectively. Conceptually, each tool is a parse tree flattener; that is, it inputs a parse tree/symbol table, extracts information from it, and outputs a token/comment stream representing the transformed program. We may remark that the Toolpack/1 installer has the option of installing these tools so that their inputs are token/comment streams – the tools then call the parser as a subroutine. Even if the tools are installed in this way, the depiction in Fig. 1 is conceptually accurate. With or without this option, a sequence of invocations of the lexer, the parser, the three transformation tools, and the formatter produce the transformed Fortran as a final result. The flow of control will be considered in Section 8.

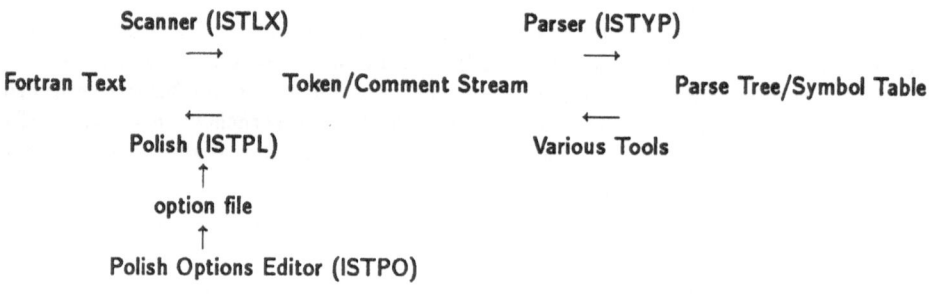

Figure 1. Transformations Among Fortran Representations

As with all Toolpack/1 tools, the transformation tools are written in Fortran that contains file-include statements as well as macro names for many constants. This source must be processed by a macro expander to produce standard Fortran that is then compiled and loaded with various Toolpack/1 libraries. The tools issue calls to subroutines and functions in these libraries, called *primitives,* to access Toolpack/1 facilities. In particular, all input/output and handling of data objects such as tokens and parse tree nodes are accomplished through calls to Toolpack/1 primitives. The Chapter "Tool Writing" provides an introduction to the primitives, data objects, names of special data objects, and other concepts in the Toolpack/1 tool-writing environment.

4.   Parse Tree Walking and Flattening

Before examining the tools in more detail, we shall first explore the facilities in Toolpack/1 employed by these and other transformation tools to extract information from the parse tree of the program being transformed and to create the token stream of the transformed program. It will be clear from the discussion that writers of transformation tools need a utility, such as the parse tree viewer ISTVT, that enables them to visualise Fortran structures as represented by subtrees of a parse tree.

As is pointed out in the Chapter "Tool Writing", the root of the parse tree is a node of type *N_ROOT* that has branches to nodes of type *N_PROGRAM* representing the program units. (We are calling node types by their macro names.) These nodes, in turn, have branches to the nodes representing the statements in the program unit. The type of each statement node and the characteristic structure of the subtree rooted at a statement node depend on the type of statement. To illustrate, Fig. 2 shows the structure of the subtree rooted at the node representing the DO statement

    20 DO 30 I = J+1,1,-1

Internally, nodes are represented as positive integers but in the figure each node is depicted by the macro name of its type. The strings associated with leaf nodes are written below the node. If the statement were not labelled, the first branch (reading left to right) from N_DO would not exist. If the incrementation parameter were the default, the last branch from N_DOSPEC would not exist.

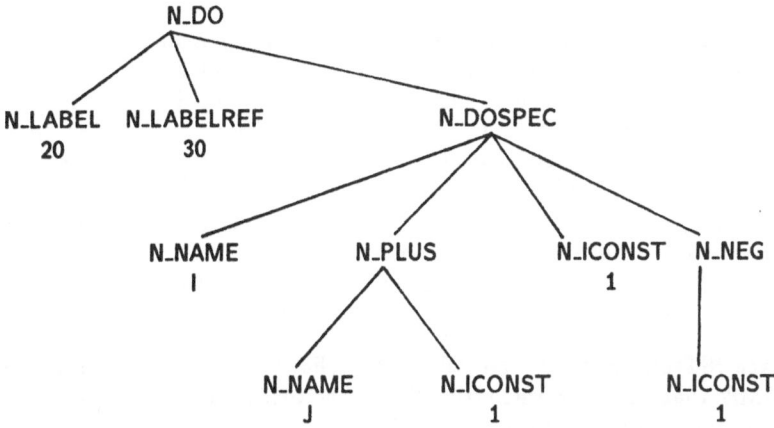

Figure 2. The Parse Representation of
20 DO 30 I = J+1, 1, -1

Given a node (represented as a positive integer) as input, there are Toolpack/1 primitives that return related nodes. The actions of these primitives can be depicted using a simple subtree consisting of three nodes, N1, N2, and N3, where N2 and N3 are on the first and second branches, respectively, reading left to right from N1. Then the primitives **ZYDOWN**, **ZYNEXT**, **ZYPREV**, and **ZYUP** behave as follows:

    N2 = ZYDOWN(N1)   N3 = ZYNEXT(N2)
    N2 = ZYPREV(N3)   N1 = ZYUP(N2)
    N1 = ZYUP(N3)

When **ZYNEXT**, **ZYPREV**, or **ZYUP** returns 0, it signals that the requested node (next, previous, or up) does not exist. When **ZYDOWN** returns 0, the requested down node does not exist and the input node is an unnamed leaf node (e.g., the node representing an END statement). When **ZYDOWN** returns a negative integer, the down node does not exist and the input node is a named leaf node with an associated string. The procedure to obtain the string as an IST string is described in the Chapter "Tool Writing".

"Tree-walking" is demonstrated by the subroutine in the Appendix. This subroutine is used by the tools to obtain certain properties of a DO loop, namely, the DO variable, the root nodes for the subtrees representing the DO parameters, and the root nodes for the subtrees representing the first and last statements in the range of the DO. Comments in the program refer to Fig. 2 so the reader can trace how the program would extract information from the subtree pictured there. The subroutine, called *DOPROP*, calls Toolpack/1 primitives and one additional subroutine, *GETSTR*, whose purpose is explained by comments in the program.

From information determined by tree walking, a transformation must determine the nature of the transformed Fortran and then write an output token/comment stream to represent it. The production of the output is accomplished by calling the token-writing primitive, **ZTOKWR**, and parse-tree-flattening primitives in the Toolpack/1 access library, as well as special flattening routines written for particular tools. The flattening primitives produce a sequence of tokens that represents the same Fortran as the subtree rooted at a given node of the parse tree. The special flattening routines produce a sequence of tokens that represents a modification of the Fortran defined by the subtree rooted at a given node. The modifications reflect the transformation being performed. Here are examples of each of these token generating routines:

(1) Let the IST string *JSTR* be "J". The following call writes a token of type *TNAME* that has associated string "J" to the token/comment stream defined by the token output channel descriptor *TKNCHN*:

```
CALL ZTOKWR(TNAME,LENGTH(JSTR),JSTR,TKNCHN)
```

The Toolpack/1 primitive **LENGTH** returns the length of an IST string.

(2) The following call to a flattening primitive writes to the output stream the sequence of tokens that represents the assignment statement defined by the subtree rooted at *NODE*:

```
CALL XYASGN(NODE,TKNCHN)
```

(3) Let the IST string *DOVAR* be "J", the IST string *ICON* be the string "3", and the Fortran expression "E" be represented as a subtree of the parse tree rooted at node *INCNOD*. The following call to a special flattening routine writes to the output stream the sequence of tokens that represents the assignment statement

defined by the subtree rooted at *NODE*, modified by replacing each occurrence of the name "J" by "J + 3*(E)".

CALL UASGU(NODE,DOVAR,ICON,INCNOD,TKNCHN)

This is the sort of substitution that would be made in an assignment statement containing the DO variable *J* when unrolling a DO loop in which the incrementation parameter is the expression "E".

5. ISTUD – The DO Loop Unrolling Tool

Using tree-walking techniques, the Toolpack/1 tool ISTUD searches a Fortran program, represented as a parse tree/symbol table, for outer DO loops, i.e., DO loops that do not appear in the range of other DOs. When it discovers such a DO loop, it first checks various conditions that the loop must satisfy to be unrolled by the tool.

All the DO loops in the input Fortran are assumed *regular*, defined as follows (see [19]): Let I be a variable of type INTEGER, F(I) be a block of Fortran statements that may depend on I, t(I) be a single Fortran statement that may depend on I, and e1, e2, e3 be expressions of type INTEGER. The Fortran DO loop

```
      DO 10 I = e1,e2,e3
            F(I)
   10       t(I)
```

is regular when

(1) The terminating statement t(I) is a CONTINUE.

(2) There are no transfers to the terminating statement from the block F(I).

(3) If the parameters are such that the loop is executed at least once, then

$$e2 = e1 + m*e3$$

for some integer m.

(4) If the loop terminates by "dropping through" the terminating statement (rather than by transferring out of the loop), then the value of the DO variable is not later used.

The assumption of regularity raises an issue that confronts tool writers, namely, the extent to which the tool can, or should, check assumptions about its input. Some conditions (e.g., regularity condition (3)) can, in a practical sense, be completely checked only at execution time and hence not by a tool that examines only a static representation of the program. Other conditions (e.g., regularity condition (4)) may, in principle, be checked by static analysis, but would cost more than requiring the user to check, especially if the check is easily made by a human or the condition satisfied by the action of another tool.

ISTUD compromises by checking conditions that help locate violations of regularity, help simplify the tool, and are easily checked by tree walking without an impractical amount of computation. Further, violations can easily be remedied by simple modifications to the input Fortran. The conditions checked by ISTUD (and the purposes they serve) are as follows:

(1) The DO loop ends on a CONTINUE (checks regularity condition (1)).

(2) The range has at least one statement besides the terminating statement (simplifies the tool).

(3) There is no transfer to the terminating label with a GO TO statement (partially checks regularity condition (2)).

(4) No inner DO loop uses the same terminating label as the DO loop being checked (simplifies the tool).

(5) Every label in the range is on a CONTINUE statement (simplifies the tool).

(6) Every statement in the range that contains the DO variable is either an assignment statement, a DO statement, or an IF statement (simplifies the tool).

If an outer DO loop passes all of the above tests, it is unrolled to the depth specified in the input by writing a sequence of tokens (including comments) representing the transformed loop. If the loop fails any of the above tests, it is not unrolled but is passed to the output token/comment stream as the sequence of tokens representing the original loop.

ISTUD determines the parameters of the original DO loop by tree walking and uses them to generate the parameters of the unrolled DO loop and the clean-up DO loop by a straightforward algorithm (see [19]). It writes a DO statement with the new parameters, using calls to ZTOKWR, and then proceeds to write the range d times where d is the unrolling depth. The first copy of the range is unchanged, while the DO variable, say "J", is replaced in copies $K = 2, ..., d$ with "J + (K-1)*(E3)", where "E3" is the incrementation parameter. Referring to the examples in Section 4, the tokens for the unchanged range are written with primitives like **XYASGN** while the other copies are written with special flattening routines like *UASGU*. There is a primitive and a corresponding special flattening routine for each allowable statement type. The primitives were models for the construction of the special flattening routines – the overall structure of each pair of corresponding routines is the same, but the special flattening routine detects the names that are to be changed and makes the required substitutions.

Assignment statements are generated in the output stream to calculate parameters of the new DO statements for the unrolled and clean-up DO statements. If the range contains labelled statements, new labels are generated for the replicated statements. Comments in the original range are reproduced in the first copy; duplicate comments are written in the clean-up loop because special comments may be compiler directives.

Comment writing primitives in the Toolpack/1 Access library utilise a file produced by ISTYP that indexes comments to statement numbers.

Following the new range, the new DO loop is terminated with a CONTINUE statement and the clean-up loop is generated using similar tree-flattening techniques. The tool then proceeds to write statements and comments from input to output until another outer DO loop is encountered.

Taking the example DO nest for matrix by vector multiplication of Section 2 as input and the unrolling depth as 4, ISTUD outputs a token stream which, when processed by ISTPL, gives

```
C *** DO-loop unrolled to depth 4 ***
      M99999 = (N- (1)+1)/ (4)
      M99998 = 1 + 4* (M99999-1)
      DO 10 J = 1,M99998,4
         DO 20 I = 1,64
            C(I) = C(I) + A(I,J)*B(J)
   20    CONTINUE
         DO 99997 I = 1,64
            C(I) = C(I) + A(I,J+1)*B(J+1)
99997    CONTINUE
         DO 99996 I = 1,64
            C(I) = C(I) + A(I,J+2)*B(J+2)
99996    CONTINUE
         DO 99995 I = 1,64
            C(I) = C(I) + A(I,J+3)*B(J+3)
99995    CONTINUE
   10 CONTINUE
      DO 99999 J = M99998 + 4,N
         DO 99994 I = 1,64
            C(I) = C(I) + A(I,J)*B(J)
99994    CONTINUE
99999 CONTINUE
```

The generated labels were not changed in the example, but ISTPL could be directed by formatting options to relabel the output Fortran. Normally, the token/comment stream output from ISTUD would not be converted to Fortran text, but would be parsed by ISTYP and the resulting parse tree/symbol table input to ISTCD.

6.   ISTCD – The DO Sequence Condensing Tool

A "sequence" of DO loops is one or more loops such that, if there are two or more, every two have the same DO variable and incrementation parameter, and every two consecutive loops are concatenated – the DO statement of the second immediately follows the terminating statement of the first. The sequences of greatest interest are the inner loops that have been generated by ISTUD in the course of unrolling an outer loop. Using tree-walking techniques, the Toolpack/1 tool ISTCD searches a Fortran program, represented as a parse tree/symbol table, for sequences of DO loops. After identifying a sequence, ISTCD determines whether it, possibly together with preceding or following statements, matches a pattern that would result from unrolling one of a particular set of paradigms of nested DO loops. If it discovers such a pattern, ISTCD condenses the sequence to a single DO loop using strategies appropriate to the pattern discovered. These include concatenating the ranges and partially expanding the loops, with suitable modification of the loop parameters. The output is a token/comment stream.

Although there are general algorithms for condensing sequences of DO loops, they produce awkward and inefficient Fortran. More seriously, they make assumptions about data independence that can only be verified at execution time; hence the transformed code is syntactically correct but unreliable.

Some insight into data dependencies can be obtained from the matrix by vector multiplication example in Section 2. Return to the point in the example where we observed that a sequence of assignment statements, divided into two groups, was equivalent to the inner DO loops in the range of an unrolled outer DO loop. We were able to move statements from the second group to the first by successively permuting pairs of statements where at no time did the left side of one appear on the right side of the other. Had we performed a permutation involving such a dependency, the result of the computation would have changed. If the original parameters and statements in the range had led to such a dependency, the loops could not have been condensed. If, say, $C(K)$ appears on the left side of an assignment statement and $C(L)$ on the right side of another assignment statement, we cannot reliably permute the statements unless we can verify that K is never equal to L. In general, it is not feasible for the program to check all possible values of K and L; rather, the program has been designed to recognise special cases for which independence has been established analytically.

The patterns recognised and transformed by ISTCD arise from paradigms for the original code. It has been shown analytically that when the outer loop of nests matching these paradigms is unrolled by ISTUD, there are no data dependencies that prevent the sequence of inner loops being condensed using the strategies employed by ISTCD. (See [19] for a typical analysis.) In one case, the absence of data dependencies hinges on conditions that the user may examine in response to messages from the tool. ISTCD may be regarded as an "expert" program that has been given a collection of problem-solving techniques. The patterns recognised and transformed by the tool may be extended as experience dictates.

At this writing, ISTCD recognises patterns derived by unrolling seven paradigms that are generalisations of DO loop nests in the linear algebra kernels in [21]. The generalisations cover 19 of the 21 kernels and are typical of DO loop nests found in basic linear algebra routines. The paradigms are defined in [15].

When ISTCD recognises a pattern from unrolling one of the paradigms, it specifies the paradigm in a message to the user and applies a built-in strategy for that paradigm. If, in examining a DO sequence, ISTCD does not recognise a pattern, it applies a general condensation algorithm and issues a warning.

The patterns are all defined in terms that can be tested by tree walking. For example, in the paradigm analyzed in [19], the DO statements for the first two loops in the DO sequence, call them

```
DO <label> I = E1,E2,E3
            and
DO <label> I = F1,F2,F3
```

must have the following properties:

```
(1) E2 = F2 (identical expressions)
(2) E3 = F3 = 1 or E3 and F3 absent
(3) For some J, either E1 = (J + a) + 1, F1 = (J + b) + 1
                      where a and b are integers, a > 0, and b > a
                  or E1 = J + 1, F1 = (J + b) + 1
                      where b is an integer, b > 0
```

Clearly, the third condition could be written more succinctly by requiring $a \geq 0$, $b > a$, but the case $a = 0$ is treated separately because tree walking tests only a specific form of an algebraic expression; in this case, the form that would be output from ISTUD.

If the unrolled code for matrix by vector multiplication shown in Section 5 were parsed and given as input to ISTCD, the output, when unscanned by ISTPL, would be

```
C *** DO-loop unrolled to depth 4 ***
      M99999 = (N- (1)+1)/ (4)
      M99998 = 1 + 4* (M99999-1)
      DO 10 J = 1,M99998,4
C *** DO loops condensed - E1 and E2 equivalence ***
C *** See ISTCD documentation for definition of PEQ conditions ***
C *** PEQ conditions are satisfied ***
      DO 55555 I = 1,64
          C(I) = C(I) + A(I,J)*B(J)
          C(I) = C(I) + A(I,J+1)*B(J+1)
          C(I) = C(I) + A(I,J+2)*B(J+2)
          C(I) = C(I) + A(I,J+3)*B(J+3)
55555     CONTINUE
```

```
   10 CONTINUE
    . DO 99999 J = M99998 + 4,N
          DO 99994 I = 1,64
              C(I) = C(I) + A(I,J)*B(J)
99994        CONTINUE
99999 CONTINUE
```

The message "Paradigm PEQ" would be output during execution to indicate the matching of one of the recognised patterns, namely, a DO loop sequence in which all the DO statements have identical parameters. This pattern is produced when the parameters of the inner DO in the original nest are independent of the outer DO variable. It is the case where the user is expected to make certain checks to assure data independence in the DO loops that were condensed. The comment

```
   C *** PEQ conditions are satisfied ***
```

indicates what should be checked. In this example, the only array that appears on the left side of a statement in any of the ranges is "C", and the only occurrence of this array on the right side of a statement has the DO variable as the index. As indicated in [15], this is enough to guarantee data independence.

Again we note that the token/comment stream output from ISTCD would not normally be converted to Fortran text, but would be parsed by ISTYP and the resulting parse tree/symbol table processed in one of several ways to be discussed in Section 8.

7.   ISTSB – The Substitution/Elimination Tool

We need some definitions from [19] to explain the action of ISTSB.

If we regard the arithmetic assignment statement $v = e$ as a definition of v, then the *dependency set* of $v = e$ is the set of variable and array references on which the definition depends. For example, the dependency set of

```
   C(I) = C(I) + A(I,J)*B(J)
```

consists of the values associated with I, C(I), J, A(I,J), and B(J). Assignment statements that contain function references are excluded, as are programs that contain EQUIVALENCE statements.

An *assignment block* is a sequence of arithmetic assignment statements, beginning either with the first assignment statement in a program unit, or the first following a non-assignment statement, and continuing until a non-assignment statement is encountered. Blocks containing assignments to subscripts of array references in the block are excluded.

Let $u = redef$ be a redefinition of $u = def$ in the assignment block

$$u = def$$
$$u_1 = v_1$$
$$\cdot$$
$$\cdot$$
$$u_i = v_i$$
$$\cdot$$
$$\cdot$$
$$u_n = v_n$$
$$u = redef$$

Suppose that, for every allowable set of fixed subscripts, none of the u_i is a reference to a member of the dependency set of $u = def$. Then the above block may be replaced by

$$u_1 = v_1 \; [u \longrightarrow def]$$
$$\cdot$$
$$u_i = v_i \; [u \longrightarrow def]$$
$$\cdot$$
$$u = redef \; [u \longrightarrow def]$$

where the first definition, $u = def$, has been eliminated and every occurrence of u on the right-hand side of any assignment statement in the block has been replaced by the expression def, suitably bracketed.

Using tree-walking techniques, the Toolpack/1 tool ISTSB searches a Fortran program, represented as a parse tree/symbol table, for assignment blocks. Having identified a block, ISTSB examines it, according to the following algorithm, and applies substitution/elimination by writing a token/comment stream.

Starting with the first statement in the block (call it statement S), the tool looks for a redefinition of its left side. Subsequent statements are examined as follows and the indicated actions taken:

(1) If the subsequent statement (call it statement T) is a redefinition, statement S is eliminated and each occurrence of the left side of statement S in subsequent statements up to and including statement T is replaced by the right side of statement S. Processing continues with a new candidate for redefinition, namely, the left side of the statement following statement T. If statement T is not a redefinition, it is checked to see if it is an assignment to a member of the dependency set of S. This is "static" checking; the tool would take A(J) and A(K) to be different, not knowing whether or not J = K. Hence, as with ISTCD, the use of ISTSB needs to be supported by the formal analysis of paradigms as in [19].

(2) If statement T is an assignment to a member of the dependency set of statement S, the tool looks ahead in the sequence for a redefinition of the left side of statement S. If it finds one, it attempts to permute the redefinition assignment statement upwards in the sequence to immediately before statement T. If the permutation conditions are satisfied (again, static checking of entries in dependency sets), the statement is moved, substitution/elimination takes place as above, and the new statement S is the statement following T.

(3) If statement T is neither a redefinition of the left side of S nor an assignment to a member of the dependency set of S, the statement following T is taken as the new statement T and processing continues. If, for a given S, no redefinition or dependency set assignment is found, the statement following S is taken as the new statement S.

(4) Processing of the assignment sequence is finished when the last statement in the sequence is reached as statement S.

If the unrolled, condensed code for matrix by vector multiplication shown in Section 6 were parsed and given as input to ISTSB and the result parsed and input a second time, the output, when unscanned by ISTPL, would be

```
C *** DO-loop unrolled to depth 4 ***
      M99999 = (N- (1)+1)/ (4)
      M99998 = 1 + 4* (M99999-1)
      DO 20 J = 1,M99998,4
C *** DO loops condensed - E1 and E2 equivalence ***
C *** See ISTCD documentation for definition of PEQ conditions ***
C *** PEQ conditions are satisfied ***
         DO 10 I = 1,64
C*** Redefinition detected - substitution/elimination applied ***
C*** Redefinition detected - substitution/elimination applied ***
            C(I) = ((((C(I)+A(I,J)*B(J))+A(I,J+1)*B(J+1))+
     +           A(I,J+2)*B(J+2)) + A(I,J+3)*B(J+3)
C*** Redefinition detected - substitution/elimination applied ***
   10       CONTINUE
   20 CONTINUE
      DO 40 J = M99998 + 4,N
         DO 30 I = 1,64
            C(I) = C(I) + A(I,J)*B(J)
   30       CONTINUE
   40 CONTINUE
```

The first two and the last two assignments in the main loop were combined on the first pass, and the resulting two statements were combined on the second pass. This is the final result, and we have instructed ISTPL to relabel the statements.

8. Operation of the Tools

Since ISTUD, ISTCD, and ISTSB are parse tree flattening tools, the parser, ISTYP, must be invoked after ISTUD to produce a parse tree/symbol table for input to ISTCD, and after ISTCD to produce a parse tree/symbol table for input to ISTSB. Moreover, whenever ISTCD or ISTSB carries out a transformation, it sets the termination status to a flag to signal which further tool sequences are to be invoked. (The termination status is set when the tool calls the termination primitive **ZQUIT**. See the Chapter "Tool Writing".)

The sequence called after ISTCD depends on the pattern of the DO loop sequence that is recognised and transformed. Most patterns are transformed iteratively, that is, a DO loop sequence is condensed into code that contains a DO loop sequence matching the same pattern or a different pattern, but having fewer DOs. ISTCD must then be repeated on the transformed code, perhaps after substitution/elimination has occurred. Thus ISTCD may set flags that call for ISTYP/ISTCD or ISTYP/ISTSB/ISTYP/ ISTCD or, if no further transformations are made, for ISTYP/ISTSB. If ISTSB carries out any substitution/elimination, it sets a flag to invoke ISTYP/ISTSB. Thus these tools make full use of the integration of the Toolpack/1 tool suite.

The command issued by the user to initiate the sequence of tool invocations depends on the user interface for a particular installation of Toolpack/1. On a Unix installation, this flow of control is managed by a shell script. The user types the command line

$$ucs < unrolling\ depth > < ISTPL\ option\ file > < Fortran\ source\ file >$$

The transform of the code in the file named $< Fortran\ source\ file >$ is written to the standard output file, and a log showing the sequence of tool invocations is written to the standard error file. The file named <ISTPL option file> is assumed to have been created using ISTPO, the Polish options editor; it governs the formatting of the transformed code by the unscanner ISTPL.

Normally, when ISTCD and ISTSB call for no further invocation sequences, the transformed program is mapped from a token/comment stream to Fortran text by ISTPL. However, if when ISTSB is invoked, no substitution/elimination transformations are possible, the presumption is that unrolling and (possibly) condensing have not served the purpose of producing Fortran better suited to a vectorising compiler. *ucs* recognises this condition and outputs the original form of the program.

Acknowledgements. The author thanks Burton Garbow and Gail Pieper for their comments on the manuscript.

Appendix:

The "Tree-Walking" Routine DOPROP

```
C -----------------------------------------------------------------
C          D O P R O P - Obtain properties of a DO loop.
C -----------------------------------------------------------------

C This source code contains macro names and must be processed with
C a macro expander, such as the Toolpack/1 installation utility
C TIEMAC, to produce compilable Fortran.

C YADEFS is a file containing various macro definitions.  YADEFS
C contains an "INCLUDE YNODES", the definitions of the macro names for
C the node types.

INCLUDE YADEFS

      SUBROUTINE DOPROP(NODE,VAR,E1,E2,E3,FIRST,LAST)

C Obtain certain properties of the DO loop whose DO statement is NODE
C on the parse tree.  The loop is assumed to end on a CONTINUE.
C Return:
C
C  VAR - the DO variable as an IST string.
C  (In Fig. 2, I is the DO variable.)
C
C  E1,E2,E3 - the root nodes for the subtrees representing the three
C  parameter expressions.  If the third parameter is the default, then
C  return E3 = 0.
C  (In Fig. 2, the three expressions are J+1, 1, and -1.)
C
C  FIRST - the root node of the subtree representing the first
C  statement in the DO range.
C
C  LAST - the root node of the subtree representing the last statement
C  in the DO range before the terminating CONTINUE.

C Arguments of DOPROP:

      INTEGER NODE,VAR(7),E1,E2,E3,FIRST,LAST

C VAR is of dimension 7 to allow for 6 characters in the DO
```

C variable plus the end-of-string character.

C Local variables:

```
      INTEGER SPTR,REFNOD,VARNOD,STLBL(6),TRMLBL(6)
```

C STLBL and TRMLBL are IST strings that hold Fortran labels, of
C maximum length 5 characters (digits) plus the end-of-string
C character.

C Toolpack/1 primitive functions:

```
      INTEGER ZYDOWN,ZYNTYP,ZYNEXT,ZYPREV,EQUAL
```

C GETSTR is not a primitive but is a subroutine to obtain the
C string associated with a named leaf node. Its call is
C CALL GETSTR(LNODE,STRING) where LNODE is a leaf node and STRING
C is an IST string that contains the string associated with the node.

```
      EXTERNAL ZYDOWN,ZYNTYP,ZYNEXT,ZYPREV,EQUAL,GETSTR
```

C The primitive ZYNTYP returns the type of a node.
C Referring to Fig. 2, NODE is the root node of type N_DO.

```
      IF (ZYNTYP(NODE) .NE. N_DO) CALL ERROR('DOPROP: Node Is Not'
     +                                  //' a DO Statement.')
```

C Get the terminating label for the DO loop.

```
      REFNOD = ZYDOWN(NODE)
```

C In Fig. 2, REFNOD is now the node of type N_LABEL, with string "20".

```
      IF (ZYNTYP(REFNOD) .EQ. N_LABEL) REFNOD = ZYNEXT(REFNOD)
```

C Now, in Fig. 2, REFNOD is the node of type N_LABELREF, with string
C "30". In general, whether the DO statement is labelled or not,
C REFNOD is now the node whose associated string is the terminating
C label for the DO loop. Hence the following call to GETSTR places
C the terminating label in the IST string TRMLBL.

```
      CALL GETSTR(REFNOD,TRMLBL)
```

C Get the DO variable.

```
      VARNOD = ZYDOWN(ZYNEXT(REFNOD))
```


C In Fig. 2, VARNOD is the node of type N_NAME whose associated string
C is "I". The following call places the DO variable in the IST string
C VAR.

```
      CALL GETSTR(VARNOD,VAR)
```

C Get the parameter nodes. If the third parameter is the default,

```
C then there is no next node after E2 and ZYNEXT(E2) returns 0.

      E1 = ZYNEXT(VARNOD)
      E2 = ZYNEXT(E1)
      E3 = ZYNEXT(E2)

C Get the first statement in the range.

      FIRST = ZYNEXT(NODE)

C Search for the last statement in the range, before the terminating
C CONTINUE by first finding the statement with label TRMLBL.  By
C assumption, this is the terminating CONTINUE statement.  The
C previous statement is then the statement sought.

C Start with the first statement in the range, found above.

      SPTR = FIRST
  100 CONTINUE

C Whatever the type of a statement, the node on the first branch is
C the label, if the statement has a label.  Here, REFNOD is the
C (potential) label node.

      REFNOD = ZYDOWN(SPTR)

C If REFNOD is a node of type N_LABEL, get its associated string and
C see if it is the terminating label.

      IF (ZYNTYP(REFNOD) .EQ. N_LABEL) THEN
         CALL GETSTR(REFNOD,STLBL)

C The Toolpack/1 primitive EQUAL compares two IST strings.

         IF (EQUAL(STLBL,TRMLBL) .EQ. yes) THEN

C We have found the statement that ends the DO loop.  LAST is
C the node of the previous statement and we have found all the
C required properties of the DO loop.

            LAST = ZYPREV(SPTR)
            RETURN
         END IF
      END IF

C One of the above conditions fails.  Try the next statement.

      SPTR = ZYNEXT(SPTR)
      GO TO 100

      END
```

The Toolpack/1 Editor and Other Fortran Modifiers

Stephen J. Hague
NAG Ltd
Oxford, U.K.

> *"A state [software system] without the means of*
> *some change is without the means of its conservation."*
>
> — Edmund Burke,
> Reflections on the Revolution in France, 1970

Abstract. In this chapter, we describe the rationale which led to the development of the Toolpack/1 editor. This editor is derived from the LBL/Unix line-based editor and contains some "Fortran 77 aware" features. Thus it can be used for both general text editing and for the modification of Fortran 77 programs. Three other Fortran 77 modifiers are also described; a pair of flexible name changers and a facility for reducing the length of non-standard names.

1.   Introduction

In the computing context, the term *editor* has traditionally meant a utility for manipulating a sequential data file consisting of records and characters. The text editor, particularly in scientific and engineering computing environments, is probably one of the most frequently accessed components of the computing system. And yet, for such a regularly used facility, the present-day text editor has a fairly low status. In general, it has not received from system designers the same attention given to other system components, and efforts to achieve some degree of portability and standardisation have yet to have any real impact. This wide diversity amongst text editors, both in terms of their design and operational characteristics, is particularly in evidence when transferring software under development from one system to another. When considering the various editors produced by the major computer manufacturers, it is clear that there has so far been little consensus about *canonical* ways of expressing

A. A. Pollicini (ed.), Using Toolpack Software Tools, 181–202.
© 1989 ECSC, EEC, EAEC, Brussels and Luxembourg.

or implementing textual changes. The pursuit of such goals was apparently not regarded as necessary or worthwhile.

There is now some evidence, however, that the importance of editors in the context of modern programming environments is being recognised. With the advent of networks, and the resulting transportation of software from one system to another, the desire for a common set of software support tools, and for a common editor in particular, soon becomes keenly felt. The notion of language-oriented editors has also been receiving increased attention in recent years. The idea of such editors is to break down the barrier between the manipulation of program source text by general purpose text editors and the processing of that text by a compiler. To varying degrees, these editors assist in the composition of programs in a syntactically correct way, and help to preserve that correctness during subsequent modification.

In the Cornell Synthesizer project [47] for example, the emphasis is on the orderly construction of programs. It has been described by its authors as providing a syntax-directed programming environment. During program construction, a top-down approach is emphasized by the provision of syntactical templates which the user is invited to instantiate. Errors in the text typed by the user within the templates are detected immediately because an in-built language parser is invoked by the editor on a phrase-by-phrase basis. Code is generated each time a template or user phrase is inserted and so execution can follow without delay. Thus program development and testing can be readily interleaved.

Syntax-directed program construction and manipulation tools in a broadly similar vein to the Cornell Synthesizer are nowadays no longer rare, but it must be said that they are not yet in widespread daily use. When editing facilities were first considered in the early days of the Toolpack project, screen-oriented program constructors modelled on the Cornell system were contemplated, but at the time the feasibility of building portable, efficient screen-based tools was doubted. Moreover, greater priority was given to the requirements of users wishing to transform existing programs rather than the development of tools to help the novice programmer to construct syntactically correct programs (though given sufficient resources, it would have been quite feasible to pursue both goals in parallel).

2.   Toolpack Editing

The strategy adopted for a Toolpack editor was as follows:

(1) first, identify or develop a transportable general purpose text editor which could provide a basic level of manipulative facilities within the Toolpack context on a variety of host computer systems,

(2) then, add Fortran-related features to make the manipulation of Fortran text easier,

(3) additionally, develop more powerful tree-based editing facilities for manipulating Fortran, in a manner which is both compatible with Toolpack conventions and can be regarded as a consistent extension of the facilities already present in the editor.

It was decided to construct a succession of editors, each progressively more powerful that its predecessors. For convenience, they were labelled editors A, B, C, D and E. The envisaged stages of development could be summarised as:

(1) a general purpose editor (editor A)

(2) Fortran-related extensions to A (editor B)

(3) an atom-based representation for editing (editor C)

(4) extension into tree editing (editor D)

(5) addition of supplementary programming facilities (editor E)

This staged development plan was conceived partly because of constraining factors at the time but also due to the desire to allow the experience gained at each stage of construction to be consolidated into the development of the next stage. The first step was to specify editor A; for that rôle the transportable version of the Unix line-based editor was chosen. Given the strong influence of Unix on Toolpack thinking, this was not a surprising choice. As a possible model of the eventual tree editing facility (stage D), the TAMPR syntax-directed transformational system [4] was borne in mind.

The current state of this development activity, as reflected in the second release of Toolpack/1, is that the Toolpack/1 editor is at stage "B+E", i.e. it incorporates Fortran-related facilities and additional programming constructs. An extended summary of the features of this editor, the Toolpack tool ISTED [28], appears in Section 3 of this chapter. Though editors C and D do not exist as distinct facilities, several Toolpack/1 tools do perform transformational operations both at the atom (token) and tree representation levels. Recognising the evolving nature of the Toolpack suite and utilising the experience gained in building transformational tools, it is anticipated that further development of the editor will take place in due course. As for currently available transformational tools, three examples of Fortran modifiers which operate on programs in a sub-source form appear in Section 4 of this chapter; ISTCN and ISTCR – a pair of versatile name changers, and ISTLS, a name standardiser.

3.   ISTED – Fortran Aware Editor

3.1.   Introduction to ISTED

ISTED is a line-based (as opposed to a screen-based) text file editor. It reads the text file to be edited into a temporary buffer ($0) for editing. All editor operations take place on the copy of the file held in the buffer, the original is not changed until a

WRIte command is given. When the file is read into the buffer all trailing blanks are deleted, no trailing blanks are ever allowed in the buffer. Files up to 5000 lines may be edited.

The temporary buffer used by the editor may exist either as a scratch file, or internally (in memory). If an external scratch file is used it may be either sequential or direct access. The location of the scratch buffer is decided by the Toolpack installer according to information given in the *"Toolpack/1 Tool Installers' Guide".*[31]

Commands to the editor have a simple and regular structure in the general style of the Unix line-based editor; zero, one or two line addresses followed by the command optionally followed by the command parameters. The structure is:

```
[line [, line]] command <parameters> [optional suffix command]
```

where '[line]' specifies a line identifier in the buffer. Every command which requires a line range specification has default values that are used if no range is specified. The line identifiers always refer to lines in the buffer, these lines are numbered consecutively starting at 1. Line identifiers may take any one of the following forms:

```
17       an integer line number
.        the current line
$        the last line in the buffer
.+n      'n' lines past the current line
.-n      'n' lines before the current line
/<pat>/  the next line past the current line that matches the pattern
         'pat'
\<pat>\  the last line before the current line that matches the pattern
         'pat'
```

Line numbers may be separated by commas or semi-colons; a semi-colon sets the current line to the previously specified identifier before the next line identifier is interpreted. This feature allows selection of the starting line for pattern (or context) searches.

Commands are entered as two or three character identifiers; these may be abbreviated to one or two characters so long as they either have no parameters or are terminated by a non-alphabetic character (e.g.: *newline*, a numeric parameter, a space etc). When more than one command starts with the one or two characters entered, then there is always a default selection; this is indicated in the individual command notes given later.

The command line is basically free format, elements of the command line may in general be separated by any number of spaces. When a specific command layout is required, then this is indicated in the individual command notes.

The editor flags errors in operation and command entry with a terse '?' or a '^?' placed where it thinks the error occurred in the command line.

3.2.   Regular Expressions

When searching for patterns, the editor allows the use of regular expressions consisting of a mixture of normal characters and metacharacters. The regular expressions are constructed according to the following rules:

(1) An ordinary character (not defined below) matches itself.

(2) A '?' matches any single character, digit or symbol.

(3) A '%' at the beginning of a regular expression matches the empty string at the beginning of a line or window (whichever starts last).

(4) A '$' at the end of a regular expression matches the null character at the end of a line or window (whichever ends first).

(5) A simple regular expression followed by a '*' matches zero or more occurrences of that regular expression (closure).

(6) A simple regular expression followed by a '+' matches one or more occurrences of that regular expression (anchored closure).

(7) A string of characters enclosed in square brackets '[ ]' matches any character in the string unless the first character is a '~' when the regular expression matches any character NOT in the string (other than *newline*). The string of characters may be abbreviated to a character range of the form a–z, 0–9, P–Y etc.

(8) The characters '<' and '>' open and close tag fields and are not part of the matching process.

(9) Any character preceded by an '@' (*escape*) including the character '@' itself matches the character without the '@' sign even if that character normally has a special meaning.

(10) The two character symbol '@n' matches the *newline* character. The two character symbol '@t' matches the *tab* character (*cntrl/I*).

(11) A ':' matches a transition between alphanumeric characters and non-alphanumeric characters or vice versa.

(12) The null regular expression represents the most recently specified regular expression.

(13) A concatenation of regular expressions is itself a regular expression.

A few restrictions apply to the use of *tabs*, macros and the *newline* '@n' symbol in regular expressions; these are noted in the *"ISTED – Fortran Aware Editor User's Guide"*.[28]

3.3.   Expressions and Conditions

There are three areas in which an expression may be entered to the editor; in a conditional expression (within a **DO** or **IF** command), in an assignment (**SET**) command or in a display (SHOw) command. Expressions can yield either a string result or a value result.

Expressions may contain user variables, constants or strings as well as binary operators and function calls. All functions are internal, and the functions currently available are given in Appendix B of the *"ISTED – Fortran Aware Editor Users' Guide"*. [28]

The operators that may be used are:

$-a$	negation (unary minus)
$+a$	unary plus
$a+b$	addition
$a-b$	subtraction
a/b	division
$a*b$	multiplication
$a**b$	exponentiation
$a\%b$	modulus

A conditional expression consists of two expressions separated by a relational operator and surrounded by parentheses; i.e.:

(<expression> <relational operator> <expression>)

Any of the expressions detailed in the previous subsection may be used for either part of the condition.

The valid relational operators are:

=	==	equals: valid for string or value comparisons
!	=!	not equals: valid for string or value comparisons
<>	><	not equals:
>		greater than: only valid for value comparisons
<		less then: only valid for value comparisons
>=	=>	greater than or equals: only valid for value comparisons
<=	=<	less than or equals: only valid for value comparisons

A conditional expression may also optionally consist of a regular expression to be matched against the current line. The pattern is delimited by '/' symbols and surrounded by parentheses, it may also optionally be preceded by a '!' symbol to indicate reversal of the pattern match sense (match a line that does not contain this pattern); i.e.:

$(/ < pattern > /)$ or $(!/ < pattern >?)$

A conditional expression may also be a '$', this is a TRUE condition if the current line is NOT the last line in the buffer.

3.4.   Simple and Global Commands

(a) Simple Commands

The following commands are provided to simplify the user interface:

(line) =	Print the line number of 'line'. Default line is current. Suffix commands 'p' and 'l' may be used.
#	Comment line. All of this line is ignored.
–	Move current line back one line and print new line.
\<return\>	Move current line forward one line and print new line.
?	Decode most recent 5 errors since last '?' or '??' command or turn off '*verbose*' switch.
??	Decode most recent 5 errors since last '?' and turn on '*verbose*' mode.
?p	Toggle the force print flag. When set to 'on' lines changed with the SUBstitute command are automatically printed (see SUBstitute). Default setting is 'off'.
?=	Toggle the line numbering switch (initial value 'off'). When the switch is 'on', then lines are displayed (e.g.: during BROwse, LISt or PRInt) preceded by their line number.

The '*verbose*' switch controls the level of information provided by the editor. When the switch is 'on' the following occur:

(1) Errors are decoded as they occur (if a decode is available).

(2) During pattern matching operations the window columns are displayed as \<start column\>, \<end column\> if any windowing other than default is in use.

(b) Global commands

There are two global commands which are used to '*filter*' the lines on which operations take place. The commands are:

```
[line [,line]] g/<regular expression>/ <command chain>
[line [,line]] x/<regular expression>/ <command chain>

[line [,line]] g:<mark character>   <command chain>
[line [,line]] x:<mark character>   <command chain>
```

The following rules apply to the use of global ISTED commands:

(1) The default line ranges are 1, $.

(2) The regular expression delimiters may be replaced by any character except ':'.

(3) The mark character must immediately follow the ':'.

(4) Mark characters are set with the MARk command.

(5) There must be a space between the closing delimiter/mark and the first character of the command chain.

The function of the global commands is to mark all lines in the specified line range which either match the expression ('g' command) or do NOT match the expression ('x' command). The commands given in the command chain are then executed on all marked lines. The command chain may consist of a single command or a sequence of commands, one per line, each (except the last) terminated by an '@'. The current limit on the command chain is 32 lines of any length.

Not all commands are legal in command chains; in particular the following commands may NOT be used:

APPend, BREak, CHAnge, DO, ENTer, INSert, MARk, REPeat and any global command.

3.5.   Summary of ISTED Commands

We now summarise the commands currently available in ISTED. For a more complete description, the reader is referred to the *"ISTED – Fortran Aware Editor Users' Guide"*. [28]

For convenience, the command summaries are grouped into four categories:

(1) editing

(2) programming

(3) display

(4) input/output

In the summaries below, commands are given in their full-name form but with their abbreviated names given in capital letters e.g. **APPend**.

(a) Editing commands in ISTED

The following commands operate on the lines of text in the editing buffer.

APPend – appends lines of text after a specified line address in the editing buffer.

CHAnge – deletes a range of lines and replaces it by new text.

COPy   – copies a range of lines (without deleting the lines) to a specified line address.

DELete – deletes a range of lines.

EXPand – expands all *tabs*, macros or both within a specified range of lines.

INSert   – similar to the **APP** command except that lines are inserted before rather than after the specified line.

JOIn   — merges a specified range of lines into a single line (provided that the merged line does not exceed the maximum line length).

MARk   — causes the specified lines to be marked with a designated character which is not actually stored in the buffer but can be used to indicate which lines are to be affected by global edits.

MOVe   — transfers a range of lines to a specified line address (similar to COPy except that the lines are deleted in their original location).

SPLit   — splits each line of a range of lines at a specified column position. If the 'f' for Fortran option is supplied, the split is performed according to the layout conventions for Fortran statements.

Most of the above commands affect only complete lines of text. There is one further editing command for the key task of intra-line editing. This command, SUBstitute, is discussed below in greater detail because of its importance within ISTED. It provides character matching and replacement facilities superior to those provided by most host system editors.

The formal ISTED specification of SUBstitute (in fact, S or SU are acceptable abbreviations, as well as SUB) is as follows:

```
(line),(line) SUB/<regular_expression>/<replacement_text>/[options]
```

In other words, within a given range of lines, SUB searches for a particular expression in each line. If it finds one or more matches in a line, it replaces the first occurrence by the supplied replacement text, unless the 'g' option is given, in which case, all occurrences within the line are replaced. Thus, for instance,

```
1,/joe/s/X/x/
```

will replace the first occurrence of 'X' by 'x' in any line containing 'X' from the first line in the buffer up to (and including) a line containing 'joe'. However, if the command were given as

```
1,/joe/s/X/x/g
```

then all instances of 'X' in the specified range would be replaced.

The power of SUB stems from the ability (derived from its Unix line editor antecedents) to use regular expressions, so that the search pattern can be expressed in quite general terms. For example, /abc[0-9]/ matches a string which starts with 'abc' followed by a digit in the range 0 to 9, /??[A-Z]*[0-9]/ matches a string starting with any two arbitrary characters, followed by none or more upper-case alphabetic characters and ending with a single digit in the range 0 to 9.

Moreover, SUB provides a tagging feature which is very useful in pattern replacement. Up to 9 expressions within the pattern matching expression can be *tagged* by surrounding them with '<' and '>'. These tagged expressions can then be re-used

repeatedly and in any order within the replacement text. The first tagged expression
is referred to as &1, the second as &2 and so on. For example:

SUB/<?>ab<?>c/&2ab&1c/

switches around the first and fourth characters, whatever they happen to be, in strings
which match '?ab?c'.

Here is another example of tagging, in this case involving character strings which will
be familiar to users of the NAG Library:

SUB/<[A-Z][0-9][0-9]><[A-Z][A-Z]>F/&2F&1/

will match six character strings only of the form

Letter digit digit Letter Letter **F**

(where *Letter* denotes an upper case alphabetic character), and will switch around
the first three and the last three characters; thus, the NAG Library routine names
E04WAF and **S17AZF**, for example, would become **WAFE04** and **AZFS17**
respectively. This tagging facility can indeed be useful when strings of a particular
pattern need to be located and modified reliably in a large body of text.

Finally, we note that the case of alphabetic characters in tagged expressions can be
changed when they are used in the replacement text:

&>n – means use the *n-th* tagged expression but with upper case characters, and

&<n – means use the *n-th* tagged expression but with lower case characters.

(b) Programming commands in ISTED

The following commands provide programming facilities within ISTED:

BREak   – causes a break out from a command loop.

CUStomize – permits the user to change the characters used to denote special
symbols, e.g.   CUS ?&   changes the arbitrary (dummy) character
from '?' to '&'.

DEClare – allows the user to define named storage areas for use during editing. All
storage areas must be declared before they are used. A variable specifier
consists of a name optionally specified by a single dimensional size, e.g.
DEC number, pi, array1(50).   Each variable or element of an array
is set to the end-of-string marker.

DEFine – defines or deletes a macro, and switches macro expansion 'on' or 'off'.
The editor allows the use of macro names to refer to frequently used text
strings. There are two types of macro: internal and those defined by
the user. The internal ones are pre-defined in the editor and consist of
abbreviations for Fortran keywords. Thus, @CX stands for COMPLEX,
@DP is an abbreviation for DOUBLE PRECISION, and so on. These

abbreviations can be displayed during an editing session by using the **SYMbols** command (described below). They are also listed in Appendix A of the *"ISTED – Fortran Aware Editor Users' Guide"*. [28]

DO   – allows the user to set up a command loop for execution of a sequence of commands with a pre-defined terminating condition. The commands are repeatedly executed until the conditional expression, which is evaluated at the beginning of each loop, is found to be *false*. The two forms of the **DO** command involve the repeated execution of a single command:

```
(line)  DO (condition) [^] command
```

or of a sequence of commands:

```
(line)  DO (condition) [^] (command
        command

              .
              .

        command)
```

If the optional '^' argument is not present in the **DO** command, the user is asked upon completion of a loop to confirm that the repeated execution is to continue. As for the '(condition)' argument, the user may alternatively supply an integer expression to specify a fixed number of iterations of the loop, or insert a '*', which signifies indefinite repetition; thus, for example,

```
DO (10) ^ S/a/b/
```

changes 10 occurrences of 'a' by 'b' in a line, and

```
/xyz/ DO (*) ( P
              /pqr/D
              S/abc/def/ )
```

causes the execution of a sequence of commands from a line starting with the string 'xyz'. The command sequence will be obeyed indefinitely except that the user is asked (because the '^' argument is not present) to confirm continuation on completion of each loop.

ECHo   – turns the echoing of command lines 'on' or 'off', as they are entered from the terminal by the user or from a script file.

FOLd   – controls the case folding switch; that is, it determines whether or not pattern matching is sensitive to the case of alphabetic characters.

HELp   – provides a summary of the commands available to the user.

IF   – allows the user to specify the conditional execution of a sequence of one or more commands. Its form is similar to that of the **DO** command.

LENgth – sets a column warning number so that if any line expands, as a result of editing, beyond that column then a message is displayed. The command can also be used to display the length of the current line.

PARse   – performs (in the currently available implementation of ISTED) a local lexical analysis of a specified sequence of lines in the editing buffer, treating those lines as one or more Fortran 77 statements. If the first and last lines to be analysed appear to be continuation lines (in the Fortran statement sense), then the range of the analysis is extended to include the Fortran statement(s) which "envelope" the specified range. The range can be specified by the user to be the "enveloping" program unit or the whole of the buffer. The user can also select the form in which the results of the analysis are displayed. Three display options are available:

Simple: – the internal Toolpack numeric codes for the Fortran tokens are displayed,

Token: – the Toolpack name of each token (as specified in the grammar table of the Toolpack lexer, ISTLX) is displayed,

Full: – the token names are displayed, and where those tokens have "values", such as a named variable or an integer constant, those values are displayed too.

In later implementations of the **PAR** command, a full syntax analysis will be available, and options to retain and manipulate the token sequence, parse tree and symbol table will be provided.

QUIt   – leaves the editor; the original file is not updated so responsibility to save the contents of the editing buffer rests with the user. If changes have been made to the buffer since the last **WRIte** operation (as described in the next subsection), the **QUIt** command queries before it is executed.

REPeat   – enables the user to look at the last ten commands entered and reselect any of them for re-execution.

SET   – allows user variables defined using the **DEClare** command to be set to the value of an expression. Its primary form is:

        `set <variable_specifier> [=] <expression>`

where the right-hand expression is considered to be a string or an integer, and the left hand side can refer to a single variable or to an array. There are two other forms, one of which resets the current line number, and the other appends the (string) result of evaluating <expression> as a line after the current line.

SHOw   – displays the current value of a user-declared variable or the result of an integer computation.

SYMbols – is used to list information about currently defined macro definitions. In selecting those macros to be listed, the user can specify all macros, internal or user-defined macros, or use a pattern; any macro with a name conforming to that pattern is listed.

TAB   – sets, controls and displays the *tab* values used by the editor which caters for ten *tab* (that is, column) positions in a line. These positions are denoted by the symbol !<*n*>, where *n* = 0 to 9. *Tabs* !0, !1 and !9 are set by default to 6, 7 and 73 respectively, and thus are convenient for entering Fortran text.

TIMe   — controls the automatic time-stamping feature of the editor. The time stamp is a comment line of the form:

*SED$ T = 11:49 - 2 OCT 1987

If a time stamp is found during a WRIte operation, and the TIMe switch is 'on', then the time stamp is updated.

(c) Display commands in ISTED

BROwse —displays a portion of the editing buffer on the terminal screen. The portion displayed can either precede, follow or surround the current line, and the number of lines displayed can be changed (the default number being 23).

HEAder —displays a column-numbering header across the top of the terminal display screen.

LISt   — displays a range of lines on the terminal screen. If control characters are detected in the lines to be LISted, they are displayed as '^X', where 'X' is an appropriate value.

PRInt   — displays a range of lines (as LISt does) but control characters are not shown.

TYPe   — allows the user to display the contents of a file without first loading it into the editor's buffer.

WINdow — sets a vertical column range for subsequent pattern matching searches. The default window is from column 1 to column 132, but can be adjusted by the user depending on the nature of the text being edited. Thus, WIN 7,72   could be useful in the context of editing Fortran statements, for example.

(d) File input/output commands in ISTED

In the following command summaries, references to files apply to both Toolpack Portable Filestore files and host system files. They can also refer to the two temporary buffers, $1 and $2, provided by the editor.

ENTer   — deletes the current contents of the editing buffer and reads in a new file. If no file name is specified, then the name of the last remembered file name (see FILe) is employed. Because the use of ENTer overwrites the contents of the buffer, its use is queried if there have been editing changes to the buffer since the last WRIte operation.

FILe   — examines or changes the name of the last remembered file.

REAd   — brings the contents of a file into the editing buffer after a specified line. The operation does not overwrite the contents of the buffer.

USE   – causes command input to be taken from a specified file, rather than from the terminal.

WRIte   – writes a specified range of lines (by default, the whole of the editing buffer) to a file.

3.6.   ISTED Internal Functions

ISTED provides a number of useful internal functions which are summarised below:

curln – returns the number of the current line.

getln – assigns a copy of the contents of a specified line into a ISTED variable.

index – finds the index (position) of a value in an array of values.

lastln – returns the line number of the last line in the editing buffer.

length – returns the number of characters in a string excluding the end-of-string marker.

string – converts its argument to a string.

substr – extracts a portion of a string into a second string variable.

union – concatenates two strings.

value – converts its argument to an integer value.

To illustrate the use of these functions, and in effect show that ISTED provides an editing programming language rather than just text editing facilities, we give below an example of an ISTED editing script, the purpose of which is to find the size of a file. Assume that we have read a file into the editing buffer in order to count the characters in the file (the *newlines* in the buffer are not included in the count of characters in the buffer). The following ISTED commands will perform such a count and can be saved in a script file for re-use. That script file may be USEd at any time that the user variables CHAR and LINE have been DEClared.

```
SET LINE = 1
SET CHAR = 0
1  DO (*) ( SET CHAR = CHAR + LENGTH(GETLN(LINE))
#             We use the GETLN and LENGTH functions to determine the
#             number of characters in the line, whose line number is
#             the value of LINE.
           SET LINE = LINE + 1
           IF (LINE > LASTLN) BRE )
#             The indefinite repetition of the DO loop is ended by
#             the BREak command when the value of LINE exceeds the
#             value of function LASTLN.
SHO = UNION("buffer size (chars):", STRING(CHAR))
SHO = UNION("file size   (chars):", STRING(CHAR + LASTLN))
#             The count of characters in the file includes newline
#             characters, the number of which is given by the value
#             of the function LASTLN.
```

Further examples of the use of the various features of ISTED, including tabbing, built-in Fortran macros, browsing, and parsing, are given in the Tutorial Example in Appendix A of the *"Toolpack/1 Release 2 Introductory Guide"*. [17]

4.   ISTCN & ISTCR – Tools for Name Changing

This section describes the use and operation of two tools that can be used for changing names within Fortran 77 program units. The two tools work at different levels of representation of Fortran and so have different capabilities.

ISTCN works at the token stream level in a manner similar to a stream editor. The input is copied to the output with any substitutions requested by the user performed. The substitutions are requested in the same form as used by the Toolpack/1 editor, ISTED, and other Toolpack/1 tools and may be selectively applied to different token types (e.g. strings, names etc).

ISTCR works at the parse tree level by transforming a symbol table. A symbol table is read in, transformed as requested by the user, then written out again.

In this description, the following metacharacters are used in the descriptions of command formats:

[...]   Optional item, the item within square brackets may be omitted.

<...>   Name of a literal item, these items are described elsewhere in the text.

4.1.   Token Changing Tool – ISTCN

(a) Parameters

The first parameter of this tool is the name of a file containing name-changing commands. There are up to four more parameters, depending upon the operation requested. Although ISTCN works at the token level, it may accept input either as source code or as a token and comment stream pair. Similarly, output may be either a token and comment stream pair or polished source code. If polished source code is required as output then a polish option file is required. If no polish option file is to be supplied, then the option file should be set to '–'. The default is for both input and output to be in token stream form. Input and output forms may be changed using control lines in the command file (see below).

ISTCN reads in a token and comment stream pair and writes modified token and comment streams. The token and comment stream pairs read and written are in ISTLX format [36]. The command file contains lists of the changes that are to be made; if the command file is selected as the standard input unit, then each command line (up to an end-of-file) will be prompted for with the 'Command: ' prompt. The format of lines in the command files is described below.

Each line of the command file may be a comment line, a control line or a change request.

Comment lines are either all blank or contain a '#' in column 1. Comment lines are ignored by ISTCN.

Input and output control lines are as follows:

       `<token`   *Set input form to token stream (default).*
       `<source`   *Set input form to source code.*
       `>token`   *Set output form to token stream (default).*
       `>source`   *Set output form to source code.*

All other lines are interpreted as change requests. A change request consists of a qualified substitution, which is based on the regular expressions described in the ISTED section of this chapter. There may be up to 256 change requests pending at one time. A change request has the following format:

       `<names> <comments> <strings> <hol> <fold> <pat> = <rep>`

`<names>` A flag to control whether this change is applied to TNAME tokens. These tokens are names, e.g. routine names, variables, common blocks etc. When a name is changed it will be checked for legality, if the new name is not-standard Fortran then a warning is issued, if the new name is neither standard Fortran nor legal for the host system, then an error is notified, but processing continues. Legality is defined by the Toolpack String Supplementary routine ZLEGAL [34]. The default is FALSE; a 'T' or 't' will set this flag TRUE.

`<comments>` A flag to control whether this change is applied to TCMMNT tokens. These tokens are comments. Default is FALSE, a 'T' or 't' will set this flag TRUE.

`<string>` A flag to control whether this change is applied to TCCNST tokens. These tokens are character constants. Default is FALSE, a 'T' or 't' will set this flag TRUE.

`<hol>` A flag to control whether this change is applied to THCNST tokens. These tokens are hollerith constants. Default is FALSE, a 'T' or 't' will set this flag TRUE.

`<fold>` A flag to indicate whether case folding is to be used in the pattern matching. Default is FALSE, a 'T' or 't' will set this flag TRUE.

`<pat>` The regular expression, as defined for ISTED, that is used for pattern matching with the name of the symbol. The pattern may not have leading or trailing spaces (except in character classes) or an embedded '='.

`<rep>` The replacement string to be used if the pattern given by `<pat>` matches the name of a token as it is read from the input stream. The string may not have leading or trailing spaces.

(b) Examples of the use of ISTCN

Some examples of the specification of change requests are given below:

```
t    t t t   f              FRED = I
```

Change the string 'FRED' to 'I' in all names, comments and strings (both character and hollerith). Note that in names (of the TNAME token type) 'FRED' would change to 'I' but 'FREDA' and 'IFRED' would not. In strings and comments ALL occurrences of 'FRED' (even 'FREDA' and 'IFRED') would be changed.

```
t ff ft I<?*> = &>1A
```

Change all names starting with 'I' or 'i' to the same string, in upper case and ending with 'A', e.g. 'IFRED' becomes 'FREDA', 'ibiLl' becomes 'BILLA'.

```
t t f f t ?* = &>0
```

Change all names and comments to upper case.

```
f t f  f t        %<?>[ ]+<?><?*> = &1 &>2&<3
```

Change all comments by removing leading spaces and converting the text to lower case, except for the first non-blank character in each comment which is to be in upper case. The comment character itself is preserved.

Note that the spacing of the flags and patterns is not significant.

4.2.   Symbol Table Changing Tool – ISTCR

(a) Parameters:

This tool has three parameters:

Parameter 1: Name of Input symbol table.
Parameter 2: Name of Output symbol table.
Parameter 3: Name of Command file.

ISTCR reads in a symbol table and writes a modified symbol table. The input and output symbol tables may have the same name. The symbol tables read and written are in ISTYP format [9]. When a name is changed it will be checked for legality, if the new name is not-standard Fortran then a warning is issued, if the new name is neither standard Fortran nor legal for the host system, then an error is notified, but processing continues. Legality is defined by the Toolpack String Supplementary routine ZLEGAL [34]. The command file contains lists of the changes that are to be made; if the command file is selected as the standard input unit, then each command line (up to an end-of-file) will be prompted for with the 'Command: ' prompt. The format of lines in the command files is described below.

Each line of the command file may be a comment, a command or a change request.

Comment lines are either all blank or contain a '#' in column 1. Comment lines are ignored by ISTCR.

Command lines contain a '%' in column 1. The following commands are available (command letter may be in either case):

%f Turn 'on' case folding, by default case folding in the regular expressions used for change requests is turned 'off', this command will turn 'on' case folding.

%l Turn 'off' listing information, by default the program lists all program units processed and all changes made, this command will cause only errors to be listed.

%q Turn 'on' querying, by default changes are made without checking with the user. If the query switch is turned 'on', all changes will be queried before being made.

%w Turn 'off' warnings. Normally a warning is issued if a resulting name is legal for the host Fortran compiler but is not standard-conforming, this option will switch 'off' the warnings.

All other lines are interpreted as change requests. A change request consists of a qualified substitution, the substitution is based on the regular expressions described in the ISTED section of this chapter. There may be up to 1000 change requests pending at any time. A change request has the following format:

[/<pu-re>/] <s_type>[:<d_types>] ([<qual>]) <pat> = <rep>

<pu-re> This is an optional regular expression to control in which program units this change is used. If no <pu-re> is given then the change is applied in all program units. Examples; '/S/' – apply only in program units called 'S'; '/S?*/' – apply in any program unit whose name starts with 'S' (e.g. S, STAND, SUM, SQUARE).

<s_type> The symbol type to be effected by the change. These types are described fully in [9]. The following symbol types may be specified (names may be preceded by 'S_' and may be abbreviated to any unique string, in either case):

COM – A common block name.
ENT – An entry point.
NAM – Names of unknown usage, e.g. variables declared but not used.
PAR – A parameter.
PRO – A procedure reference.
PU – Program unit names for programs, block data, functions and subroutines. All program units have names, unnamed main programs are called '$MAIN', unnamed block data is called '$BLOCKDATA'.
STA – A statement function.
VAR – A variable.

<d_types> These are optional data type qualifiers, this allows you to specify only INTEGER procedures or REAL variables etc. If this qualifier is not used then all data types are deemed to match. These types are described in [9]. More than one qualifier may be used, each is a single character (in either case) as follows:

    B – block data
    P – program unit or subroutine
    I – integer
    R – real
    L – logical
    X – complex
    D – double precision
    C – character
    G – generic, for standard intrinsic functions

<qual> These qualifiers specify the usage a symbol must be put to in order to be classed as a match. These qualifiers equate to the symbol attributes defined in [9]. If no qualifiers are specified then any appearance of the symbol is deemed to match. The following qualifiers are available (these may be abbreviated to any unique string in either case):

    ARG – used as an argument.
    ASS – appears in an ASSIGN statement.
    COM – appears in a COMMON statement.
    DAT – appears in a DATA statement.
    DUM – is a dummy argument or formal parameter to the program unit.
    EQU – appears in an EQUIVALENCE statement.
    EXP – appears in an expression.
    EXT – appears in an EXTERNAL statement.
    FUN – used as a function.
    IND – used as a DO loop index.
    INT – appears in an INTRINSIC statement.
    REA – appears in a READ input list.
    SET – set by an assignment statement.
    SF   – dummy argument or formal parameter of a statement function.
    STA – standard intrinsic.
    SUB – used as a subroutine.
    USE – The symbol has been used in some way, this is equivalent to all of the
           following (DUM, ASS, REA, DAT, SF, EQU, ARG, FUN, SUB, IND).

<pat> The regular expression that is used for pattern matching against the name of the symbol. The pattern may not have leading or trailing spaces (except in character classes) or an embedded '='.

<rep> The replacement string to be used if the pattern matches the name of the symbol. The string may not have leading or trailing spaces.

Note, not all qualifiers or data types are relevant to all symbol types. In particular, neither data types nor qualifiers can be used with COMMON symbol types (i.e. common block names).

Once the replacement has been done the resulting name is checked to see if it is both legal (as defined by a routine in ISTCR) and unique within the program unit. If either of these tests fails then an error occurs, if the command file was specified as the standard input unit then the user will be prompted for an alternative name, otherwise the program stops.

(b) Examples of the use of ISTCR

Some examples of the specification of change requests are given below:

```
VAR:I() FRED = I
```

In all program units, change integer variables named 'FRED' to 'I', whatever their usage.

```
VAR() FRED = I
```

In all program units, change all variables named 'FRED' to 'I', whatever their usage or data type.

```
/S?*/ VAR:IR(INDEX) I = LOOP
```

In all program units whose name starts with 'S', change integer and real variables which are used as DO loop indices from 'I' to 'LOOP'. Note, this may not have quite the desired effect if implicit typing is used as any real variables would become integers (this may, of course, be the intention in this instance).

```
VAR:I (USE) I<?*> = &1A
```

In all program units, change used integer variables whose names start with 'I' so that the names end with 'A', e.g. 'IFRED' becomes 'FREDA', 'IBILL' becomes 'BILLA' etc.

```
PU:I () <????>?<?> = I&1&2
ENT:I () <????>?<?> = I&1&2
PRO:I (FUN) <????>?<?> = I&1&2
```

Change all integer functions and integer function entry points which have 6 character names so that the names start with 'I', dropping the fifth character. Change all references to those functions as well. Note, no check is made as to the uniqueness of a new PU name, checks are only made on changes within a PU.

```
PU:P () <????>?<?> = Q&1&2
ENT:P () <????>?<?> = Q&1&2
PRO (SUB) <????>?<?> = Q&1&2
```

Change all subroutines and subroutine entry points which have 6 character names so that the names start with 'Q', dropping the fifth character. Change all occurrences of those names in CALL statements as well. Note, no check is made as to the uniqueness of a new PU name, checks are only made on changes within a PU. Note also that this will change the name of the main program, if it contains 6 characters.

```
PU:b () <?><?><?><?><?><?> = &6&5&4&3&2&1
```

Reverse the names of all block data program units whose names have 6 characters. Note, no check is made as to the uniqueness of the new PU name.

```
CO () <???><???> = &2&1
```

Change the names of all common blocks that currently have 6 character names. Reverse the first and last groups of 3 characters.

4.3.   Long-Name Changing Tool – ISTLS

The Toolpack long-name processor, ISTLS, detects long names in a token stream and substitutes Fortran 77 standard-conforming names (that is, no more than six characters in length) supplied by the user in a name conversion file or in response to prompts from the tool. The substitute names supplied by prompting are saved in an updated version of the conversion file which can be used in subsequent runs. Thus, the user can either employ ISTLS to make a single standardising pass to produce a standard-conforming program (with respect to the length of names), or develop and maintain her or his program in its long-name form and treat ISTLS as a preprocessor to the host system Fortran 77 compiler.

ISTLS has six parameters, the rôle of which is as follows:

Parameter 1: the name of the file containing the token stream of the program to be standardised.

Parameter 2: the name of the corresponding comment stream file (these two files having been produced by the Toolpack lexer, ISTLX).

Parameter 3: the name of the file to contain the revised token stream output by ISTLS. This file can then be processed by other tools, including the polishing tool, ISTPL, which restores it to source form.

Parameter 4: the name of the output comment stream file.

Parameter 5: the name of the file holding the name conversion instructions. The file consists of a list of names, one per line, with each long name on an odd-numbered line and the corresponding standard-conforming name on the following even-numbered line. It will normally be created and updated by running ISTLS and responding to prompts.

Parameter 6: the name of a log file, in which a record of changes made during ISTLS processing is kept.

For an example of the use of ISTLS, with input and output files, the reader is referred to the *"ISTLS Users' Guide"*. [16]

Documentation and Non-Fortran Tools

Ian C. Hounam,
NAG Ltd
Oxford, U.K.

"Any software package is only as good as its user documentation."

— Brian Ford,
Software Transfer and Sharing – An Overview,
in *D.T. Muxworthy [ed], Programming for Software Sharing,* 1983

Abstract. This chapter is concerned with the Toolpack/1 documentation and report generating tools. The Toolpack/1 text formatter is described in detail. Two text comparison tools and a version control system are summarised. Two tools to aid in the generation of reports and a comparison tool for numeric data files complete this section.

1.   Introduction

The previous chapters have dealt with the way in which Toolpack/1 can provide direct assistance to the Fortran programmer in the analysis and transformation of his programs.

This chapter deals with the ways in which Toolpack/1 can provide aids in the documentation of programs, text formatting and a general interface to the Toolpack/1 suite.

A. A. Pollicini (ed.), Using Toolpack Software Tools, 203–215.
© 1989 ECSC, EEC, EAEC, Brussels and Luxembourg.

2.   ISTRF

This tool provides a general text formatter. It gives similar functions to the Unix "roff" [44] and the VAX/VMS "runoff" [20] programs. In order to illustrate the function of this program some of the basic commands will be discussed and a short example document prepared. This is not an exhaustive treatment of the tool and the reader is recommended to consult the ISTRF User's Manual for further details [18].

The Toolpack/1 base tape is supplied with all the tool documentation in ISTRF format. This allows the installer to make changes to the format, for example the page size, to suit local requirements.

ISTRF takes as input a file containing text, in random length lines, and formatting commands. It produces as output a page formatted document with a given line and page length. Lines of text can be filled to the given length and optionally right justified.

Commands are given as two letters following the control character (the default is .) The following example uses the commands '.sp' to specify (one) space and '.ti 3' for a temporary indent of 3 characters. The input file is:

```
.sp
.ti 3
ISTRF
will, by default, put all the text
in these random length lines
into a right
justified paragraph.
```

The output from ISTRF, if the right margin is set to 25, will be:

```
   ISTRF    will,    by
default, put all the text
in these random  length
lines   into    a   right
justified paragraph.
```

This illustrates the way in which ISTRF commands are distinguished from text. Any line that starts with a dot (the default command character) is analysed as a command, if there is a syntax error a warning is issued.

2.1. Page Formatting

The commands that define the page format are illustrated in Figure 1 to show the page position that they refer to. The syntax for these commands is:

.fo '*left*'*centre*'*right*'   – Footer titles
.he '*left*'*centre*'*right*'   – Header titles
.in n   – Indent by n characters, i.e. left margin at n+1.
.m1 n   – Margin above header is n lines.
.m2 n   – Margin below header is n lines.
.m3 n   – Margin above footer is n lines.
.m4 n   – Margin below footer is n lines.
.pl n   – n Lines per page.
.rm n   – Right margin is n.

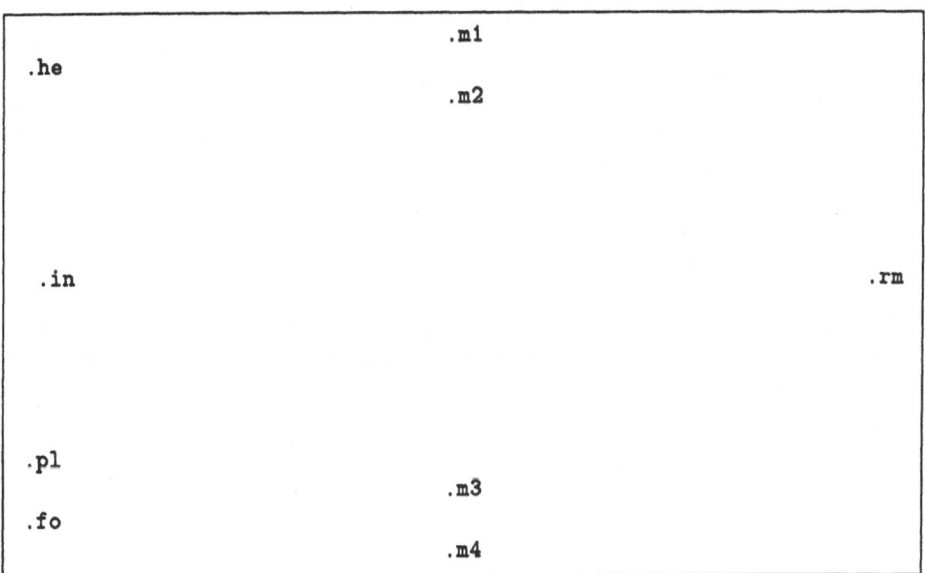

Figure 1. Page Format Commands

2.2. Headings

The command '.ce' causes the next line of input text to be centred. '.bd' causes the next line to be printed in bold face, and '.ul' underlines the next line. These can be used together, for example,

```
.ce
.ul
.bd
1. Introduction
```

will produce as output:

1. Introduction

The TIE implementation may provide a method for bolding and underlining by backspacing or double printing. Alternatively, in-line printer codes may be specified to toggle these operations on and off using the commands (.ib, .nb, .iu and .nu), see *"ISTRF – Text Formatter"* [18] for details. In some cases the text may require postprocessing, for example, to drive a laser printer. Extra commands to set up and control the printer may be inserted automatically by such a tool.

2.3. Laying Out Text

Two paragraph commands have already been encountered in the simple example in Section 2. These and other commands for formatting paragraphs and laying out tables are given below:

```
.br    – Break, start a new line.
.fi    – Fill output lines (default).
.nf    – Stop filling output lines.
.ju    – Right justify lines (default).
.nj    – Stop justifying.
.ti n  – Indent the next line by n+1 characters.
.ne n  – Start a new page if there are less than n lines left on the current page.
.bp    – Start a new page.
.sp n  – Break and leave n blank lines.
```

These commands are all illustrated with the following example. Figure 2 is the input file and Figure 3 the output text.

```
.pl 72
.rm 50
.fo 'ISTRF Example''Page #'
.ce
.bd
EXAMPLE OF ISTRF TEXT FORMATTING
.sp 3
.ce
by
.sp
.ce
.ul
Ian Hounam
.sp 2
.ti 3
This text will be filled into lines 72 characters
long and right justified because the default is for these
to be on.  The paragraph may be broken.
.br
This will appear on a new line.
.in 15
A hanging paragraph can be produced by using the '.in' command.
This will cause all text up to the next '.in' command to be
indented to column 16.  '.in 0' will return the left margin
to column 1.
.in 0
.sp
.ti 3
Where it is necessary to output text exactly as it is typed in,
the commands '.nj' and '.nf' should be used to switch off
text justifying and line filling.  This is used, for example,
to produce a fixed format for tables or references.
.nf
.nj
.sp
REFERENCES

1.  The ISTRF User Guide.
    Toolpack/1, Release 2.1, NAG Publication NP1318.
.sp
.ju
.fi
The commands '.ju' and '.fi' turn the normal paragraph
formatting back on again.
```

Figure 2. Example ISTRF input file

The output from ISTRF looks like this:

```
              EXAMPLE OF ISTRF TEXT FORMATTING


                          by

                     Ian Hounam

     This text will be both filled into lines 72
characters long and right justified because the
default is for these to be on. The  paragraph  may
be broken.
This will appear on a new line.
                A hanging paragraph can be produced
                by using the '.in' command. This
                will cause all text up to the  next
                '.in' command to be indented to
                column 16. '.in 0' will return  the
                left margin to column 1.

     Where it is necessary to output text exactly as
it is typed in, the commands '.nj' and '.nf'
should be used to switch off text  justifying  and
line filling. This is used, for example, to
produce a fixed format for tables or references.

REFERENCES

1. The ISTRF User Guide.
   Toolpack/1, Release 2.1, NAG Publication NP1318.

The commands '.ju' and '.fi' turn the  normal
paragraph formatting back on again.
```

Figure 3. Output from ISTRF

Lines which start with a space are not filled or justified, provided they fit within the margins. This provides an easy way of inputing a line or two to be printed in a fixed format.

2.4.   Macros

The user can define macros of commonly used combinations of commands for example to create section headers or to provide the page format for commonly used paper sizes. Such macros may be introduced as a separate file or may be defined in the main document file.

The macro definition starts with the command '.de xx', where xx is the name (case sensitive) of the macro. Any ISTRF commands may follow ending with the '.en' command. The macro is invoked by the command line '.xx'. An example of a macro, called A4, to set up the page size for A4 paper is:

```
.de A4
.pl 72
.rm 72
.m1 6
.m2 2
.m3 2
.m4 6
.en
```

This is one of the Toolpack/1 Technical Documentation Macros and is used in all the Toolpack/1 documentation distributed with the base tape.

3.   ISTCB

This tool is a text file comparison program. Its inputs are two versions of a document and its output is the document with added ISTRF commands to generate change and delete bars. These are characters that are printed on the right margin to mark changed and deleted text.

The new ISTRF commands are:

.bb – Start output of change bar
.eb – End output of change bar
.db – Output a delete bar.

The area where changes have been made to the original text is marked by the default change bar character |. Because the text is being filled into lines the tool may not place the change bar exactly at the change, normally there is an overlap before or after the section with changes. The deleted section is printed out, marked by a #.

```
.rm 50
This is an example document to illustrate the
operation of ISTCB. Some changes will be made
to this text.
.br
ISTCB will be run to compare the two files and
the output from ISTRF will be printed out.
.br
This line will be deleted in the modified file.
```

Figure 4. Example Input Text for ISTCB

```
.rm 50
This is an example document to illustrate the
operation of the tool ISTCB. Some changes will be made
to this text.
.br
This line is inserted.
.br
ISTCB will be run to compare the two files and
the output from ISTRF will be printed out.
```

Figure 5. Example Modified Text for ISTCB

```
.rm 50
This is an example document to illustrate the
.bb
operation of the tool ISTCB. Some changes will be made
.eb
to this text.
.br
.bb
This line is inserted.
.eb
.br
ISTCB will be run to compare the two files and
the output from ISTRF will be printed out.
.db
```

Figure 6. The Output File from ISTCB

```
This  is  an  example  document  to illustrate the   |
operation of the tool ISTCB. Some changes will  be   |
made to this text.                                    |
This line is inserted.                                |
ISTCB will be run to compare the two files and the
output from ISTRF will be printed out.               #
```

Figure 7. The Output File from ISTRF

4.   ISTTD

ISTTD is a general text comparison tool. Its input is two versions of a document. The output is a list of the differences.

There are four options that control the manner of the comparison.

LMARG=n – Sets the left margin for the window to be searched.

RMARG=n – Sets the right margin for the window to be searched.

Only characters columns between LMARG and RMARG are used in the comparison.

BLANKS   – Leading and trailing blanks are ignored, and multiple blanks are reduced to single blanks. (Default – off)

FOLDING   – Case is ignored in the comparison. (Default – off)

Each line that differs is listed with an indication of whether it has been inserted in or deleted from the reference file. Any changed lines are listed as being deleted and inserted.

5.   ISTVC

This tool maintains information about changes made to any formatted sequential file. ISTVC maintains a separate Version Control File (VCF) for the file being monitored. As a new version is added ISTVC compares the new version with the latest version held and writes the differences to the VCF. Earlier versions of the file can be recovered by specifying either the version number or the date and time. When recreating a file the tool recognises the macro V# which is replaced by the version number. This facility can be used to include the version number automatically in, for example, a comment line near the start of each piece of code.

ISTVC is not intended to support large software projects but can be useful in providing a change control system for small to medium size programs where the host system does

not provide one. Instead of storing copies of each stage of evolution of a program, one
file contains the latest version of the program or text. A second file contains the
commands necessary to edit this file back to its state at any previous time. This file
also contains user supplied comments, normally the reason for the change, embedded
with each change.

6.   ISTDX

This tool will extract all marked comments from a program and insert them in a
document.

Blocks of comments are marked for extraction using a Source Embedded Directive
(SED), this is a special format of comment that is interpreted as a command by a
Toolpack/1 tool [13]. ISTDX SEDs are:

    *dx=on   – To switch on comment extraction.

    *dx=off   – To switch off comment extraction.

Only comments which start with * will be extracted. The * will be removed on output.

Example DX-1

Input program (start):

```
        PROGRAM DEMO
        DO 10 I=1,10
   10   PRINT *,A,B,C
*$dx$=on
* These comment lines will be extracted
* by the tool ISTDX
*$dx$=off
* These comments will not be extracted
* because extraction is switched off
*$dx$=on
C These will not be extracted because they
C use the letter C as the comment character
```

Output text:

```
        These comment lines will be extracted
        by the tool ISTDX.
```

7.   ISTAL

ISTAL is a static and dynamic analysis report generator. It may be run interactively or may be used to insert reports into an ISTRF type document. ISTAL normally requires that other tools, ISTYP or ISTAN or both, are run first. ISTAL then generates reports based on processing the output of these tools.

7.1.   Annotated Listing

For a program unit or units that have been instrumented by ISTAN [11], ISTAL can produce a listing annotated with the execution frequencies of all segments of code. If the instrumented program contains assertions to ISTAN to check if a logical expression is true, the failure counts are incorporated. This information is derived from the run time statistics file generated by running the instrumented program.

7.2.   Segment Execution Frequencies

The segment execution frequencies, as above, may be tabulated. Each segment is a piece of straight line code, that is a series of statements with no transfer of control. A table may be produced showing the number of times each segment is executed for all program units.

Alternatively this information may be combined with static analysis statistics (program statement numbers etc.) to give a table summarising the total segment execution in each program unit.

Additionally, ISTAL can also report on any segments of unexecuted code.

7.3.   Statement Type Counts

ISTAL can produce tables of the numbers of the various types of statement in each unit. The count can be a static one where the symbol table from ISTYP is used to give the numbers of each type. Alternatively the statistics file from the instrumented program can be used to give the dynamic count of the execution of each statement type.

7.4.   Documentation Extraction

In a similar mode of operation to ISTDX, ISTAL can be used to extract marked blocks of comments from a source file. This mode is turned on and off by the same SEDs as

ISTDX. (see above). ISTAL, though will extract all comments whether they start with C, c or *. The comment character is removed and the comment lines inserted in the output file.

7.5.   Modes of Operation

If ISTAL is used interactively, that is both input and output is to the user's terminal, then the tool takes commands in the form of:

    command = file_name   or:
    command = program_unit_name

e.g. To produce the dynamic statement execution count for the program unit SQUARE:

    dynamic = SQUARE

This will cause the table to be listed at the terminal. Instead of the program unit name a wild card may be specified, for full details please see *"ISTAL Users' Guide"* [37].

An alternative method of using ISTAL is to specify as input an ISTRF file containing ISTAL commands of the form:

    .al command = file_name

A command of this form will cause the ISTAL reports to be inserted in the document with appropriate ISTRF commands. An example of the output tables generated by this tool is shown in the Chapter "Workshop on Fortran Analysis".

8.   ISTDC

ISTDC is the Toolpack/1 data comparison tool. It is designed mainly for comparing files of numeric values though it is also capable of comparing files with embedded text.

This tool will examine versions of a data file and report differences. ISTDC first tries to interpret the characters from the input files as numbers. These are held as double precision variables, comparison is then made to within a specified tolerance (default 1E-6). If the string can not be interpreted as a number then the comparison is made as a string. The tool treats spaces, by default, as delimiters only. The number of spaces between items is not significant in the comparison.

Example DC-1

The following files would be found identical if the tolerance for numerical comparisons were set to 1E-6.

File 1
1.1234561
2.123456E3
1.2 2.3   3.4

File 2
1.1234562
2123.457
1.2   2.3 3.4

9.   Conclusion

This chapter has described some of the ways that Toolpack/1 tools can help the Fortran programmer prepare documentation. A portable text formatting system has been described that has applications for all aspects of document preparation.

Supporting Coding Conventions

Aurelio A. Pollicini
Joint Research Centre
Ispra, Italy

*"Audi, Israël, et observa ut facias quæ præcipit
Dominus, et bene sit tibi, et multipliceris amplius,
sicut pollicitus est Dominus Deus patrum tuorum
tibi terra lacte et melle manentem."*

— Deuteronomy, 17,5

Abstract. In the spirit that norms and laws are meant to be beneficial to the target community, a series of conventions addressed to Fortran programmers is discussed with the intent of promoting the development of clearer and more understandable Fortran 77 code. The way of applying these conventions automatically, by using Toolpack tools, is shown in practice.

1.   Introduction

Within the context of modern software development methodologies, the coding aspect is neither the more critical issue, nor highly costly in term of resources required. Nevertheless, a careful use of programming languages is worth of our attention. Compliance to international standards, when they exist, as in the case of the programming language Fortran [1], is essential. Additionally, in areas where the common practice has not yet received support from any standardizing activity is worth adopting general or even local conventions.

We may refer to conventions which are relevant to the coding phase at two levels: *programming conventions* and *coding conventions*. Since the Fortran Language is still widely used for scientific computation, in the aim of complementing and integrating the usefulness of the ISO standard, a number of projects on conventions specific to Fortran 77 were undertaken. In particular, from 1982 to 1984, an international collaboration between Électricité de France, CERN and the Commission

A. A. Pollicini (ed.), Using Toolpack Software Tools, 217–231.

of the European Communities aimed at producing a common set of rules for programming conventions. That ambitious goal was not completely met, however, a high commonality was reached. While the conventions adopted by CERN [42] contain about 30% of site-specific requirements, the conventions adopted by EDF [24] and those proposed as an internal draft for the Commission [14], in addition to matching the other 70% of CERN rules, reached an almost full agreement on functionalities, though the editorial forms were very different. Many of the choices presented in this chapter are inspired by that agreement.

Programming conventions step over two phase of the software development cycle, that is detailed design and coding. They also imply some decisions on what statements have to be used. Therefore, without a restructuring tool they cannot be applied to existing software. Therefore, the presentation only deals with coding conventions and specifically with those rules concerned with how to group statements and identify that groups, as well as with the order and the form in which statements are expressed.

2.   Coding Conventions: Elements and Implementation

Following a top-down approach we will consider the order of statements, the segmentation of the program into a declaration-part and an execution-part, the transformations applicable to both parts. Finally, transformations at the line level will be considered.

2.1.   Main Rules Concerning the Order of Statements

Some basic requirements for the order of statements within a program unit are defined by the standard for the Programming Language Fortran. However, the high flexibility allowed for specification statements as well as the freedom concerning the location of FORMAT and DATA statements may downgrade the visibility of some program attributes and its global clarity.

Additional requirements on statement order derive from general considerations about programming languages design. Thus, the stress on the visibility over entities is put, in decreasing order, on symbolic constants, global variables, local variables, procedures. Then the sequence of executable statements gives a compact view of the underlying algorithm. Moreover, DATA statements and FORMAT statements should be gathered in two groups of statements. The former group is moved after the variable entities, since compile-time assignment adds a value attribute to the variables, and the latter one is moved to the end of the program unit for an easy reference.

Consequently, we consider the first set of rules.

"After the heading statement of a program unit, insert a packet of appropriate type

statements and PARAMETER statements, so that the symbolic names of constants are first explicitly typed and then defined."

"Within each program unit, group all DATA statements together and insert them after the specification statements."

"Within each program unit, group all FORMAT statement together and insert them before the END statement."

The option **mode** of the declaration standardizer ISTDS, described in detail in the Chapter "Fortran 77 Transformers", will cope with the first two rules as we will see in Section 3 below, while the third rule above is implemented by setting the option MOVEF=.TRUE. in the Polish option file.

2.2. Rules for Structuring the Declaration-part

The fundamental requirement for the declarative sections is that all entities should be declared. Further requirements concern the segmentation into sections ordered according to the levels of visibility. At a lower level, a sound order has to be defined for the allocation of the different data types as a function of the storage units required. ISTDS provides an implementation for most of these requirements, and in particular for the following rules.

"Associate an explicit type to every symbolic name used to identify either a constant or a variable entity or a function subprogram or a statement function."

"Avoid the use of the IMPLICIT statement and provide explicit typing instead."

"For all variable entities make a distinction between scalars and arrays."

"Avoid the use of the DIMENSION statement and specify the array declarators in the relevant type statements so that a visual relationship is established between every array and its type and size attributes."

"After the packet of statements defining the symbolic constants, group into separate sections the specification statements used to declare, where appropriate, the dummy arguments, the entities in Common and the local entities."

"Declare the standard Data Types in the decreasing order of storage units requirements, that is COMPLEX, DOUBLE PRECISION, REAL, INTEGER, LOGICAL and CHARACTER."

"For character entities and procedures, append the length specifier to the keyword CHARACTER rather than to the entity identifier."

"After the declaration sections concerning variable entities, declare all the procedures in separate sections: external functions, subroutines, intrinsic function and statement functions."

"Group all Common blocks together, by using one COMMON statement per block."

"Group all SAVE statements together as the last section of specification statements."

Remark that conformance to the standard requirement of using separate Common blocks for CHARACTER and non-character entities, only relies on the user's discipline. Moreover, ISTDS does not alter entity order within Common blocks. Therefore, within Common blocks, entity allocation according to the decreasing storage requirement mentioned above, also depends on the user's discipline.

2.3. Rules for Tidying the Execution-part

The transformations applied to executable statements aim at improving the clarity of the execution-part. They deal with order of labels, separation of nested DO-loops and indentation of blocks in agreement with the following rules.

"Enforce the semantic distinction between statement labels and FORMAT labels by using two separate label ranges."

"For each range, use equally spaced increasing sequences of label numbers."

"Use the CONTINUE statement as termination statement of DO-loops."

"For nested DO-loops sharing the same termination statement (and label) enforce DO range separation by using a distinct CONTINUE as termination statement for each DO-loop."

"Make use of indentation for giving evidence to block structure and nesting."

An implementation of the rules above is provided by suitable Polish options, as shown below.

The following set of option values fulfils the requirements on labels:

```
RLBSTM  = .TRUE.
  SLBINI =     10
  SLBINC =     10
RLBFMT  = .TRUE.
  FLBINI =   9000
  FLBINC =     10
```

The following set of option values fulfils the requirements on DO-loops and indentation:

```
DOCONI  = .TRUE.
INDDOC  = .TRUE.
  INDDO =      2
  INDIF =      2
  INDCON =    -1
```

Additionally, the Polish option section BLANK_LINES provides parameters for controlling code segmentation using blank lines.

2.4. Line Cosmetics

The Polish option file provides supplementary options for breaking statements (see section LINE_BREAK), token spacing (see sections SPACING1 and SPACING2) and line bounding.

LMARGS, RMARGS, LABELC, LABELF may be used for changing the boundaries of some fields of the source statements. Comment lines may be indented and/or bounded by combinations of the options INDCMT, LMARGC, RMARGC. Conventional line identification is also provided by the following option values:

```
SEQRQD  = .TRUE.
  SEQINI =    10
  SEQINC =    10
  SEQDIG =     4
  SEQFIL =   '0'
```

3. ISTDS Selected Options

The most important option for our purposes is the operating mode. The default is **mode=rebuild_declaratives** which causes reformatting of the declaration-part into well defined declarative sections. We may either keep the default value or confirm it explicitly for better clarity.

Concerning data type declaration, ISTDS allows us to select different forms:

data_type=keywords   uses DOUBLE COMPLEX and DOUBLE PRECISION;

data_type=length_specifiers   appends the length in bytes to the keywords COMPLEX, REAL, INTEGER, LOGICAL;

data_type=not_double_complex   corresponds to "keywords" except for long complex which are typed as COMPLEX*16. There is no default for this option, therefore, if not specified no change will apply to the original use of type keywords.

According to our purpose of conforming to the standard we reject the use of length specifiers. Since the remaining two forms only differ for a non-standard data type, either might be chosen. However, the last one seems to put more evidence on a use which violate the standard.

Therefore, because of this higher visibility we may prefer it.

The order of data type statements is fully parametrized by the option:

order=<sequence>.

The default is: DOUBLE COMPLEX, DOUBLE PRECISION, COMPLEX, REAL, INTEGER, LOGICAL, CHARACTER. Apart from the first type which is non defined in standard Fortran, there is only one slight difference between the above order and the one specified by the rules of the previous section. That is DOUBLE PRECISION and COMPLEX are inverted. Although their relative position is immaterial since both data types have the same storage mapping, we modify the order as follows, simply for aesthetic reasons:

order=8,4,5,2,1,3,6

By default, in each declarative section all character entities are declared by using a single CHARACTER statement. Since we require the different length specifiers to follow the keyword rather than the entity name, we specify the option chlbrk which produces the requested separation.

Concerning Common blocks, there are two options — ardicb and ictwcb — which may be turned on for controlling a different distribution of the specification statements that refer to Common entities. Both options are off by default. Since we agree with Common entities after dummy argument entities, we do not specify these options.

Since all declarative sections are separated by header comments of the form:

C   .. Section Name ..

a comment in the same style is inserted before the first executable statement by specifying the option exehdr.

Finally we may take profit of the option remove_unused_names if we want to prune our source program of any dead token. The default is "no", but we may choose either "yes" or, alternatively, "log" which has the same effect as "yes" and additionally issues a message for each removed name. The latter seems better because it makes the user aware of any intervened deletion of unused entities.

By summarizing, the options string suggested for our purposes is:

data_type=not_double_complex order=8,4,5,2,1,3,6 chlbrk exehdr
remove_unused_names=log

4.   The Preliminary ISTPO Session

We create a new option file on PFS with name "workshop.opt" for which only the "Basic" and "Common" option sections will have some default changed. Subsequently, different parametrizations may be actually tested during the practical session.

Remark that prompts and output issued by the tool ISTPO start from column two, so that input lines are those beginning in the leftmost position.

```
 Option file name:
workshop.opt
 : New file:

 PO>
menu basic

 Menu: BASIC                            Next: COMMON

 SEQRQD: Add sequence numbers                = .FALSE.

 RLBFMT: Relabel FORMAT statements           = .FALSE.
 RLBSTM: Relabel executable statements       = .FALSE.
 MOVEF : Move FORMAT statements to end       = .FALSE.

 DOCONI: End each DO-loop on a CONTINUE       = .FALSE.
 IOTHCO: Put CONTINUE on each labelled stmt  = .FALSE.

 TRACE : Display progress messages           = .FALSE.

 ERRCMT: Insert error messages as comments   = .FALSE.

 CONCHR: Continuation character                = '+'


 Type ? for help

 PO>
seqrqd.true.

 PO>
rlbfmt=.true.

 PO>
rlbstm=.true.

 PO>
movef=.true.

 PO>
doconi=.true.

 PO>
next

 Menu: COMMON                          Next: UNCOMMON

 SEQINI: Initial Sequence Number             = 1
 SEQINC: Sequence Number Increment           = 1
 SEQDIG: Number of digits in seq number      = 4
 SEQFIL: Fill character for seq number       = ' '

 SLBINI: Initial statement label             = 10
 SLBINC: Statement label increment           = 10
```

```
FLBINI: Initial FORMAT label                = 9000
FLBINC: FORMAT label increment              = 10

INDDO : Indentation within a DO-loop        = 4
INDIF : Indentation within a block-IF       = 4
INDCON: Continuation line indentation       = -2
INDCMT: Indent comments as statements       = .FALSE.
INDDOC: Indent DO- CONTINUE's with body     = .FALSE.

VLEN  : Padding before "=" for variables = 0

 PO>
seqini=10

 PO>
seqinc=10

 PO>
seqfil='0'

 PO>
inddo=2

 PO>
indif=2

 PO>
indcon=-1

 PO>
inddoc=.true.

 PO>
exit
```

5.   The Transformation Chain

For the sake of completeness we run at least once the full chain: ISTLX, ISTYP, ISTDS, ISTPL.

For experiencing different transformations we will be able to start by the parsed view of the sample program stored on PFS and rerun repeatedly the chain ISTDS/ISTPL only.

The tool chain may be run either by using a host command file or by means of ISTCE commands as it is explained in Section 2 of the Chapter "Toolpack Invocation Techniques". The arguments to be passed to the tools consist of input and output file names, with the only exception of the seventh argument passed to ISTDS, which tells

the tool the processing options chosen for the run. The following option string is used
in this example:

```
data=not_double_complex order=8,4,5,2,1,3,6 chlbrk exehdr remove=log
```

Remark that the line above specifies processing options equivalent to those shown at
the end of Section 3. Indeed, any unambiguous abbreviation of an option name is valid.

The polish option file input to ISTPL is the one updated by the option editing
commands shown in the previous section.

6. Sample Program

For giving a flavour of how some particular piece of the Toolpack/1 software need to
be customized on specific systems, we have chosen the version of the SPLIT utility for
IBM MVS/TSO environment. Since the initial form of the source is not as relevant as
the final presentation, we show in this section only the transformed source.

```
      PROGRAM SPLIT                                              SPLIO0010
C-------------------------------------------------------------   SPLIO0020
C    TOOLPACK/1     Release: 2.1                                 SPLIO0030
C----Ispra Frontend----NB. Replace homonymous  file "SPLIT"     SPLIO0040
C                                                                SPLIO0050
C                                                                SPLIO0060
C    THE OBJECT OF THIS PROGRAM IS TIED TO THE USE OF CHARACTER  SPLIO0070
C    VARIABLES FOR USE IN THE F77 OPEN STATEMENT.                SPLIO0080
C    WHAT THIS PROGRAM DOES IS SPLIT UP A FILE AT CERTAIN        SPLIO0090
C    MARKS IN THE FILE.  ON THE SAME LINE AS THE SPECIAL         SPLIO0100
C    MARK, THE NAME OF THE FILE THAT THE NEXT CHUNK OF CODE      SPLIO0110
C    IS TO BE PUT IN APPEARS.                                    SPLIO0120
C    THIS FILE NAME IS USED IN THE OPEN STATEMENT,               SPLIO0130
C    AND THE CODE FROM THAT POINT ON IS WRITTEN TO THE NAMED FILE. SPLIO0140
C                                                                SPLIO0150
C                                                                SPLIO0160
C                                                                SPLIO0170
```

```
C     .. Parameters ..                                        SPLI0180
      INTEGER LXT                                             SPLI0190
      PARAMETER (LXT=80)                                      SPLI0200
C     ..                                                      SPLI0210
C     .. Local Scalars ..                                     SPLI0220
      INTEGER FREE,RC1,RC2,STDIN,STDOUT,STRLEN,UNIT           SPLI0230
      LOGICAL START                                           SPLI0240
      CHARACTER BLANK                                         SPLI0250
      CHARACTER*7 VSFILE                                      SPLI0260
      CHARACTER*8 DIRHDR                                      SPLI0270
      CHARACTER*44 HOSTNM                                     SPLI0280
      CHARACTER*(LXT) FILENM,LINE                             SPLI0290
C     ..                                                      SPLI0300
C     .. External Functions ..                                SPLI0310
      LOGICAL KEYLIN                                          SPLI0320
      CHARACTER*8 QJRCDH                                      SPLI0330
      CHARACTER*120 QJRCDR                                    SPLI0340
      EXTERNAL KEYLIN,QJRCDH,QJRCDR                           SPLI0350
C     ..                                                      SPLI0360
C     .. External Subroutines ..                              SPLI0370
      EXTERNAL OUTPT,QJRCHF,TSOBOX                            SPLI0380
C     ..                                                      SPLI0390
C     .. Data statements ..                                   SPLI0400
C                                                             SPLI0410
      DATA BLANK/' '/                                         SPLI0420
      DATA START/.TRUE./                                      SPLI0430
      DATA STRLEN/120/                                        SPLI0440
C     ..                                                      SPLI0450
C     .. Executable Statements ..                             SPLI0460
C**********************************************************    SPLI0470
C     HOST SYSTEM DEPENDENT DATA STATEMENT......              SPLI0480
C         SET STDOUT TO THE STANDARD OUTPUT UNIT NUMBER       SPLI0490
C         SET STDIN  TO THE STANDARD INPUT  UNIT NUMBER       SPLI0500
C                                                             SPLI0510
      STDOUT = 6                                              SPLI0520
      STDIN = 5                                               SPLI0530
      FREE = 10                                               SPLI0540
C**********************************************************    SPLI0550
                                                              SPLI0560
      UNIT = STDOUT                                           SPLI0570
```

```
                                                             SPLI0580
10 CONTINUE                                                  SPLI0590
   READ (STDIN,FMT='(A)',END=20) LINE                        SPLI0600
   IF (LINE.NE.BLANK) THEN                                   SPLI0610
     IF (KEYLIN(LINE,FILENM)) THEN                           SPLI0620
       CALL OUTPT(LINE,UNIT)                                 SPLI0630
       IF (UNIT.EQ.STDOUT) THEN                              SPLI0640
         UNIT = FREE                                         SPLI0650
                                                             SPLI 660

       ELSE                                                  SPLI0670
         CLOSE (UNIT)                                        SPLI0680
         CALL TSOBOX(QJRCDR('FF',VSFILE,BLANK),STRLEN,RC1,RC2)  SPLI0690
       END IF                                                SPLI0700
                                                             SPLI0710

       CALL QJRCHF(FILENM,DIRHDR,VSFILE,HOSTNM)              SPLI0720
       CALL TSOBOX(QJRCDR('AO',VSFILE,HOSTNM),STRLEN,RC1,RC2)  SPLI0730
                                                             SPLI0740

       FILENM = VSFILE                                       SPLI0750
       OPEN (UNIT,FILE=FILENM,STATUS='UNKNOWN')              SPLI0760
       REWIND (UNIT)                                         SPLI0770
                                                             SPLI 780

     ELSE                                                    SPLI0790
       IF (UNIT.EQ.STDOUT) THEN                              SPLI0800
         WRITE (UNIT,FMT='(1X,A)') LINE                      SPLI0810
         IF (START) THEN                                     SPLI0820
           DIRHDR = QJRCDH(LINE)                             SPLI0830
           START = .FALSE.                                   SPLI0840
         END IF                                              SPLI0850
                                                             SPLI 860

       ELSE                                                  SPLI0870
         CALL OUTPT(LINE,UNIT)                               SPLI0880
       END IF                                                SPLI0890
                                                             SPLI 900

     END IF                                                  SPLI0910
                                                             SPLI 920

   END IF                                                    SPLI0930
                                                             SPLI 940

   GO TO 10                                                  SPLI0950
                                                             SPLI0960
20 CLOSE (UNIT)                                              SPLI0970
   STOP                                                      SPLI0980
                                                             SPLI 990

   END                                                       SPLI1000
```

```
      LOGICAL FUNCTION KEYLIN(LINE,FILE)                        KEYL0010
C-------------------------------------------------             KEYL0020
C     special program unit separator is looked for             KEYL0030
C-------------------------------------------------             KEYL0040
C                                                              KEYL0050
C     THIS LOOKS AT THE LINE OF CODE FOR A SPECIAL MARK        KEYL0060
C     IF THE MARK IS FOUND, IT LOOKS FURTHER FOR              KEYL0070
C     THE FILE NAME THAT THE NEXT CHUNK OF SHOULD BE NAMED     KEYL0080
C                                                              KEYL0090
C                                                              KEYL0100
C                                                              KEYL0110
C                                                              KEYL0120
C     .. Scalar Arguments ..                                   KEYL0130
      CHARACTER*(*) FILE,LINE                                   KEYL0140
C     ..                                                       KEYL0150
C     .. Scalars in Common ..                                  KEYL0160
      LOGICAL ISCMNT,ISEOF                                      KEYL0170
C     ..                                                       KEYL0180
C     .. Local Scalars ..                                      KEYL0190
      INTEGER PNTR                                              KEYL0200
      CHARACTER BLANK,COMFL,COMFU,COMR                          KEYL0210
C     ..                                                       KEYL0220
C     .. Intrinsic Functions ..                                KEYL0230
      INTRINSIC LEN                                             KEYL0240
C     ..                                                       KEYL0250
C     .. Common blocks ..                                      KEYL0260
      COMMON /COLUM1/ISCMNT,ISEOF                               KEYL0270
C     ..                                                       KEYL0280
C     .. Save statement ..                                     KEYL0290
      SAVE /COLUM1/                                             KEYL0300
C     ..                                                       KEYL0310
C     .. Data statements ..                                    KEYL0320
                                                               KEYL0330
                                                               KEYL0340
      DATA COMFL/'c'/                                           KEYL0350
      DATA COMFU/'C'/                                           KEYL0360
      DATA COMR/'*'/                                            KEYL0370
      DATA BLANK/' '/                                           KEYL0380
C     ..                                                       KEYL0390
```

```
C     .. Executable Statements ..                              KEYL0400
                                                               KEYL0410

      KEYLIN = .FALSE.                                          KEYL0420
                                                               KEYL0430

      ISCMNT = .FALSE.                                          KEYL0440
      ISEOF = .FALSE.                                           KEYL0450
C     ONLY LOOK AT COMMENTS THAT START IN COLUMN ONE (EFL,RATFOR)  KEYL0460
                                                               KEYL0470

      IF ((LINE(1:1).EQ.COMFL) .OR. (LINE(1:1).EQ.COMFU) .OR.  KEYL0480
     +    (LINE(1:1).EQ.COMR)) THEN                            KEYL0490
                                                               KEYL0500

C       LOOK FOR THE REST OF THE KEY- '$$$SPLIT$$$'            KEYL0510
                                                               KEYL0520

        IF (LINE(2:12).EQ.'$$$SPLIT$$$') THEN                  KEYL0530
          PNTR = 13                                            KEYL0540
   10     IF (LINE(PNTR:PNTR).EQ.BLANK) THEN                   KEYL0550
            PNTR = PNTR + 1                                    KEYL0560
            IF (PNTR.GT.LEN(LINE)) RETURN                      KEYL0570
            GO TO 10                                           KEYL0580
                                                               KEYL 590

          END IF                                               KEYL0600
                                                               KEYL 610

          FILE = LINE(PNTR:)                                   KEYL0620
          KEYLIN = .TRUE.                                      KEYL0630
          ISEOF = .TRUE.                                       KEYL0640
                                                               KEYL 650

        ELSE                                                   KEYL0660
          ISCMNT = .TRUE.                                      KEYL0670
        END IF                                                 KEYL0680
                                                               KEYL 690

      END IF                                                   KEYL0700
                                                               KEYL 710

      RETURN                                                   KEYL0720
                                                               KEYL 730

      END                                                      KEYL0740
```

```
      SUBROUTINE OUTPT(LINE,UNIT)                              OUTP0010
C-----------------------------------------------------------  OUTP0020
C     source line handler (move to individual P.U. file)      OUTP0030
C-----------------------------------------------------------  OUTP0040
C                                                              OUTP0050
C                                                              OUTP0060
C                                                              OUTP0070
C                                                              OUTP0080
C                                                              OUTP0090
C     .. Parameters ..                                         OUTP0100
      INTEGER LXT                                              OUTP0110
      PARAMETER (LXT=80)                                       OUTP0120
C     ..                                                       OUTP0130
C     .. Scalar Arguments ..                                   OUTP0140
      INTEGER UNIT                                             OUTP0150
      CHARACTER*(*) LINE                                       OUTP0160
C     ..                                                       OUTP0170
C     .. Scalars in Common ..                                  OUTP0180
      LOGICAL ISCMNT,ISEOF                                     OUTP0190
C     ..                                                       OUTP0200
C     .. Local Scalars ..                                      OUTP0210
      INTEGER COUNT,L                                          OUTP0220
      CHARACTER BLANK                                          OUTP0230
C     ..                                                       OUTP0240
C     .. Local Arrays ..                                       OUTP0250
      CHARACTER*(LXT) LQUEUE(256),SQUEUE(256)                  OUTP0260
C     ..                                                       OUTP0270
C     .. External Functions ..                                 OUTP0280
      LOGICAL QJRCHS                                           OUTP0290
      EXTERNAL QJRCHS                                          OUTP0300
C     ..                                                       OUTP0310
C     .. Common blocks ..                                      OUTP0320
      COMMON /COLUM1/ISCMNT,ISEOF                              OUTP0330
C     ..                                                       OUTP0340
C     .. Save statement ..                                     OUTP0350
      SAVE /COLUM1/,COUNT,LQUEUE,SQUEUE                        OUTP0360
C     ..                                                       OUTP0370
C     .. Data statements ..                                    OUTP0380
C                                                              OUTP0390
      DATA COUNT/0/                                            OUTP0400
      DATA BLANK/' '/                                          OUTP0410
C     ..                                                       OUTP0420
```

```
C     .. Executable Statements ..                                 OUTP0430
C                                                                  OUTP0440
      IF (LINE.EQ.BLANK) RETURN                                    OUTP0450
      IF (ISCMNT) THEN                                             OUTP0460
        COUNT = COUNT + 1                                          OUTP0470
        IF (COUNT.LE.256) THEN                                     OUTP0480
          LQUEUE(COUNT) = LINE                                     OUTP0490
                                                                   OUTP 500
        ELSE IF (COUNT.LE.512) THEN                                OUTP0510
          SQUEUE(COUNT-256) = LINE                                 OUTP0520
                                                                   OUTP 530
        ELSE                                                       OUTP0540
          PRINT *,'REMOVED COMMENT: ',LINE                         OUTP0550
        END IF                                                     OUTP0560
                                                                   OUTP 570
      ELSE IF (.NOT.ISEOF) THEN                                    OUTP0580
        IF (QJRCHS(LINE)) THEN                                     OUTP0590
          WRITE (UNIT,FMT=9000) LINE                               OUTP0600
                                                                   OUTP 610
        ELSE                                                       OUTP0620
          IF (COUNT.GT.0) THEN                                     OUTP0630
            DO 10 L = 1,COUNT                                      OUTP0640
              IF (L.LE.256) THEN                                   OUTP0650
                WRITE (UNIT,FMT=9000) LQUEUE(L)                    OUTP0660
                                                                   OUTP 670
              ELSE                                                 OUTP0680
                WRITE (UNIT,FMT=9000) SQUEUE(L-256)                OUTP0690
              END IF                                               OUTP0700
                                                                   OUTP 710
   10       CONTINUE                                               OUTP0720
            COUNT = 0                                              OUTP0730
          END IF                                                   OUTP0740
                                                                   OUTP 750
          WRITE (UNIT,FMT=9000) LINE                               OUTP0760
        END IF                                                     OUTP0770
                                                                   OUTP 780
      END IF                                                       OUTP0790
                                                                   OUTP 800
 9000 FORMAT (A)                                                   OUTP 810
      END                                                          OUTP0820
```

Part III
A Multi-dimensional Toolpack View

*"If I have seen further it is by standing
on the shoulders of giants."*

— Isaac Newton,
   Letter to Robert Hooke, 5th Feb. 1675

General Advice on Installing Toolpack/1

Ian C. Hounam,
NAG Ltd
Oxford, U.K.

"Therefore I want to address my closing remarks to the system programmers and to the machine designers who produce the systems that the rest of us must work with. Please, give us tools that are a pleasure to use, especially for our routine assignments, instead of providing something we have to fight with. Please, give us tools that encourage us to write better programs, by enhancing our pleasure when we do so."

— Donald E. Knuth,
Computer Programming as an Art,
Turing Award Lecture, CACM 17, December 1974

Abstract. This chapter guides the Toolpack/1 installer through the stages of selection of the appropriate version of TIE. For some installations this is available in a ready to use form, for others the installer must modify the supplied code. For the latter the paper covers the selection of a TIE regime and the installation of the simplest regime. The code that must be modified to do this is dealt with in some detail. The installation of the supplementary libraries, and finally the tools, is explained.

1. Introduction

Toolpack/1 is normally supplied in Fortran 77 source code form. Toolpack/1 is considered to be transportable across a very wide range of machines. On most machines there should be no great difficulty in installing Toolpack/1, but on machines that do not support a standard conforming Fortran 77 compiler more programmer work will be necessary in the installation process.

Some mention is made here of the various formats in which Toolpack/1 is distributed. Full details of these may be found in Appendix A.

A. A. Pollicini (ed.), Using Toolpack Software Tools, 235–249.
© *1989 ECSC, EEC, EAEC, Brussels and Luxembourg.*

As has been explained in the Chapter "Structure of Toolpack/1 Software", the Toolpack/1 suite is arranged in such a way that all possibly machine dependent code is contained in TIE – Tool Interface to the Environment. The first part of this chapter will concentrate on installing TIE. The way in which one possible TIE regime may be implemented is explained. Examples are given of the type of change that is necessary in order to provide the machine dependent functions of the Toolpack Virtual Machine.

The second part of this chapter deals with installing the supplementary libraries. This presents no great difficulty because these are written in portable Fortran. Finally the installation of the tools themselves is dealt with.

2.   Which version of TIE?

There are two machine specific versions of Toolpack/1 available for VAX/VMS and Unix. These are the easiest ways to implement Toolpack/1. The next level of TIE implementation comes from contributed TIECODE installations. A case study regarding one such implementation is reported in Appendix B. Finally for completely new installations the base tape contains the basic TIECODE installation which requires modification by the installer.

2.1.   VAX/VMS

As a new service at Release 2 all the Toolpack/1 tools are available on a tape containing executable programs for DEC VAX/VMS version 4.2 or later. It is highly recommended that VAX/VMS sites that have not already installed Toolpack/1 Release 1 use this tape for their initial installation.

If Toolpack/1 is to be installed on VAX/VMS using the base tape either,
 (1) because the executable program tape is not being used or
 (2) because a re-installation is necessary after a bug fix,

the implemetation of TIE for VAX/VMS – TIEVMS should be used.

TIEVMS is written in Pascal and Fortran 77 and provides an efficient implementation of TIE for this range of machines. An executable version of TIEVMS is supplied on the VAX/VMS specific tape for sites that do not have a Pascal compiler. In fact, on this tape, the executable forms of the tools are linked with this version of TIE.

2.2.   Unix Systems

NAG also distributes a Unix system specific version of Toolpack/1. This consists of a tar tape written with a suitable directory format of all the source programs. A "Make" file is provided to carry out all compilation and binding operations. This simplifies the initial installation procedure.

TIEC, an implementation of TIE written in C for Unix machines, should be used if the Unix specific tape, which also contains and uses TIEC, is not available.

2.3.   Contributed tapes

As Toolpack/1 is public domain (public access) software, co-operation among sites is encouraged. From the first release of Toolpack/1 there is available a set of six contributed TIECODE installations for the following machines and operating systems:

1)	IBM	MVS/TSO
2)	IBM	VM/CMS
3)	Cyber	NOS/2
4)	Cyber	NOS/VE
5)	Harris	VOS
6)	DEC/10	TOPS/10

The installation of contributed TIECODE is outwith the scope of this chapter, but, in general these tapes are fully documented. For these systems and similar ones, these tapes offer the best starting point for the installer.

2.4.   Other Machines

For machines other than those dealt with above the implementation of TIE that should be used is TIECODE. TIECODE is written entirely in Fortran 77, but requires some machine specific functions to be provided. Section 5 of this chapter is devoted to an explanation of how one regime of TIECODE can be implemented.

3.   The Toolpack/1 Base Tape

This is the most general Toolpack/1 distribution service. All of the tools and all three versions of TIE are on this tape in source code form. The entire set of documentation is provided in ISTRF format and certain key documents are already page formatted for immediate printing.

3.1.   The SPLIT Utility

The Toolpack/1 base tape files are mostly made up of concatenated files separated by
split markers in the form of special comment lines. These specify the file name to which
any following lines are written out by the SPLIT program until another split marker
is encountered.

Example Split-1

PROG.FILE

```
C$$$SPLIT$$$ PROG.MAC
       PROGRAM PROG
       ...
       END
C$$$SPLIT$$$ SUBR.MAC
       SUBROUTINE SUBR...
       ...
       END
```

PROG.MAC

```
       PROGRAM PROG
       ...
       END
```

SUBR.MAC

```
       SUBROUTINE SUBR
       ...
       END
```

3.2.   The TIEMAC Utility

The TIEMAC program provides the following facilitites for preprocessing Fortran
programs:

(1) Macro substitution.
    Definitions of the form:

    DEFINE (macro_name,value)

    may be made, either in the program itself or include files. All occurrences of
    "macro_name" in the program will be substituted by "value". TIEMAC leaves
    comments unchanged.

(2) Include file inclusion.
    A portable include file capability is provided. Statements of the form:

    INCLUDE include_file_name

    cause input to be switched from standard input to the named file. Include files
    may be nested.

(3) File name redefinition.

A statement of the form:

`FILE (filename1 , filename2`

causes any occurrence of "filename1" on an include statement to be replaced by "filename2". For example,

`FILE (yadefs , /toolpack/tie/access/yadefs`

shows how a filename may be replaced by the fully qualified filename. The FILE statement does not have a closing right parenthesis to allow for file names that contain ")".

Example Tiemac-1

PROG.MAC

```
        PROGRAM PROG
INCLUDE DEFS
INCLUDE COMN
        DATA MAX/max_values/
        ...
        END
```

PROG.FTN

```
        PROGRAM PROG
        COMMON /X/A,B,C
        DATA MAX/1000/
        ...
        END
```

DEFS

```
DEFINE(max_values , 1000)
```

COMN

```
        COMMON /X/A,B,C
```

In Toolpack/1 TIEMAC is used to:

(1) Include COMMON blocks. Only one definition of each COMMON block exists. An INCLUDE statement is used to insert the block in the appropriate place in each routine where it is used. This ensures consistency of all COMMON block usage, and avoids the difficulty to trace bugs that can occur when COMMON definitions differ between routines.

(2) Perform macro substitution. Where the type of an entity in a Toolpack/1 program is associated with a numeric value the program may use a macro name for that numeric value that says what the type is. Finally in preprocessing with TIEMAC the numeric value is substituted. Macro definitions can either be made in the

program code itself, or in include files. Hence a file of macro definitions may be included in communicating programs ensuring that a value has the same meaning in both.

(3) Include PARAMETER statements. One file of PARAMETER statements can be inserted in all program units that use them. This is one way that consistency of PARAMETER statements can be ensured.

3.3.   Installing SPLIT and TIEMAC

Both SPLIT and TIEMAC are written in Fortran 77 and in many circumstances are ready to run, but they may require some customisation before use.

- **Both Programs**

The I/O unit numbers need to be checked to ensure that they are available for use:

(1) Standard output (STDOUT) (and for TIEMAC standard error (STDERR)) are set to unit 7, this may require to be changed (e.g. to 6 on some systems).

(2) Standard input (STDIN) is set to unit 5, if this is not suitable it should be changed.

All READ, WRITE, OPEN and FORMAT statements should be checked for host compatibility.

- **SPLIT only**

The unit used to OPEN the output split file is assigned in the statement:

    FREE = 10

If this is not suitable it should be changed.

Filenames of the form ABCDEF.XYZ are used on Toolpack/1 split markers. If these are not suitable, then code should be added to the subroutine KEYLIN to modify these in some way to conform to the host file naming rules.

- **TIEMAC only**

TIEMAC, as set up, will use units 11 to 13 for input include files. These may be changed if necessary.

4.   Installing TIECODE

This section guides the user through the choice of a TIE regime (see Chapter "TIE Specifications") and the selection of files from the base tape to create these.

4.1.   **The TIECODE Files**

The six TIECODE files are:

TIE.FILE   – Include Files containing COMMON blocks and a definition file for the TIECODE routines.

COMMON.FILE – The TIE routines common to all operating regimes.

HOST.MAC   – I/O routines for host file store operation only.

PFS.FILE   – I/O routines for both host and portable file store access.

DIRECT.MAC   – Flow of control routine dummies for regimes that do not require one tool to schedule another.

EMBED.MAC   – Flow of control routines for tool scheduling.

Not all of these files need to be recovered from the tape in the first instance. The next section will make it clear which files are necessary.

4.2.   **Stand-Alone Regime**

Tools are called by, and return control to, the host operating system. All files are host files whether or not the file name is preceded by # (the host file identifier). This is the easiest TIE regime to install.

Files Required:

```
TIE.FILE
COMMON.FILE
HOST.MAC
DIRECT.MAC
```

4.3.   **Stand-Atop Regime**

Again no tool scheduling is supported by this regime. The portable file store is implemented in this regime.

Files Required:

```
TIE.FILE
COMMON.FILE
PFS.FILE
DIRECT.MAC
```

4.4.   Stand-Astride Regime

Tool scheduling is available but no PFS is implemented.

Files Required:

```
TIE.FILE
COMMON.FILE
HOST.MAC
EMBED.MAC
```

4.5.   Embedded Regime

This is the full implementation of TIE including both the portable file store and tool scheduling capabilities. ISTCE and ISTCI are fully operational only under this regime.

Files Required:

```
TIE.FILE
COMMON.FILE
PFS.FILE
EMBED.MAC
```

4.6.   Choosing a TIE regime

For an initial implementation of TIE it is recommended that the Stand-Alone regime is chosen. No tool scheduling is involved in this regime which removes a part of TIE that is entirely machine dependent and hence requires the addition of machine specific code to TIE. No provision is made to use the portable file store.

Although this is the minimum TIE implementation, it may be all that is required. If the following two conditions are met then there will be little point in adding any more functionality to TIE:

(1) All Toolpack/1 users are familiar with the host machine operating system, hence there is no need to provide an isolated environment for the tool user like, for example, ISTCE.

(2) The host machine provides a directory based file system.

5.   Installing Stand-Alone TIECODE

This section gives an overview of the steps involved in installing TIECODE in the stand-alone regime. Full details of this procedure can be found in the *"TIE Installers' Guide"* [30].

5.1.   File 1 – TIE.FILE

This file should be split to give XDEFIN and a number of files with name starting XC...

XDEFIN is an *include file* which is automatically included with each TIECODE routine. It contains:

(1) TIEMAC FILE commands to give the full pathname for the XC... files.

e.g. FILE(XCDEVI /TOOLPACK/TIE/XCDEVI

These should be changed to specify the directory where the XC... files have been placed, in the host machine file name format.

(2) A number of macro definitions that specify some host operating system characteristics. These should be carefully checked and the appropriate values substituted, where necessary. Some of these macro values are calculated from others so if a change is made then the calculated macros should also be checked.

Examples of Macros:

DEFINE(bpi , 32)   *bits per integer*

DEFINE(funithost , 10)   *first host unit number*

DEFINE(maxfiles , 13)   *maximum number of files*

5.2.   File 2 – COMMON.FILE

This file contains the TIE routines that are common to all operating regimes. When split there are over 60 individual .MAC files, most containing one routine.

Before going any further a number of points need to be checked:

(1) Unit numbers for I/O.

(2) READ, WRITE, FORMAT, INQUIRE and CLOSE statements must be checked for host system compatibility.

(3) The IST character set definition should be checked for characters that have a special meaning on input or on output. If any, an alternative means of representing

such a character is provided by providing a two character sequence where an alternative character, preceded by a backquote ('), represents the special character. This is done by altering the DATA statements that control the character conversion on input and output. The internal representation of the tab character should also be set here.

Customisation

Some routines require to be customised because their function can not be provided in portable Fortran 77. An example of this is the routine ZTIME. This routine returns the date and time and requires the addition of call to the appropriate host system routine(s) and the addition of any code necessary to change the format to that required.

In other cases the customisation necessary would be carried out to improve the appearance of the output or to improve efficiency.

ZPRMPT is the routine that outputs the prompt string. This uses a Fortran WRITE and FORMAT statement. As supplied, a new line is output after the prompt. If this can be suppressed then the FORMAT statement can be altered, or a call made to a system prompt routine, as appropriate.

ZLOGIC contains 8 logical functions, 5 of which must be modified (ZIAND, ZINOT, ZIOR, ZLLS, ZLRS) These can be replaced by either:

(1) host system calls,

(2) local intrinsic functions or,

(3) installer written assembler functions.

ZFIELD, ZBYTE and ZCBYTE are coded in Fortran 77 and need not be changed.

5.3.   File 3 – HOST.MAC

This file contains two functional routines and a number of dummy routines that are irrelevant in the Stand-Alone regime.

Only one of the functional routines needs to be checked. This is the routine XLIONM, which forces all file names to be in the host file store irrespective of whether they are marked by the host file character (#).

If file names of the form:

```
/DIR1/DIR2/FILE.TYP
```

are acceptable to the host system then no changes need to be made here. If the host file system has a directory structure then code should be added here to convert the file name to the correct format. For example, if host file names have the following format, the above file name could be converted to:

```
[DIR1.DIR2]FILE.TYP
```

If there is no directory structure file store then, if it is possible, the Toolpack/1 file names should be mapped onto unique host filenames. This almost certainly implies a great restriction on the format and length of Toolpack file names, and probably represents only a temporary measure until the PFS is installed.

5.4.   File 4 – DIRECT.MAC

This file contains two dummy versions of the flow of control routines for chaining and spawning tools. These dummy subroutines simply terminate the tool operation and ensure return of control to the operating system. There should be no changes to make to this file.

5.5.   Preparing the TIE library

The following steps have now to be carried out on the four TIE files:

(1) SPLIT, in the cases of TIE.FILE and COMMON.FILE.
(2) Preprocessing by the TIEMAC utility, except TIE.FILE.
(3) Compilation by a standard conforming Fortran 77 compiler.

Now all the object files can be placed in a selective load library, if the host computer supports this. This simplifies the commands to the linker, loader or binder when the tools and supplementary libraries are being installed.

The test program in SUPPORT.FILE should be recovered from tape, compiled and linked with the TIE library. Although this program refers to both the HFS and PFS, in this regime all files are HFS files. In the case of a problem arising, the test program will help to narrow down the search for the bug. Although it is only a limited test of the TIE facilities it should be run at this stage to ensure that the basic fuctions work.

6.   Supplementary Libraries

The base tape contains files for all 4 supplementary libraries:

(1) `ACCESS.FILE`
(2) `STRING.FILE`
(3) `TABLES.FILE`
(4) `WINDOW.FILE`

The last of these, the WINDOW supplementary library is used only by one tool – ISTVT. There is no need to install this library until all the other required tools are up and running.

As the parser ISTYP requires the first three libraries, only the most minimum of implemetations can be run with less than these three. The TABLES supplementary library should present no problem in installation therefore it is recommended that all three are installed allowing all the tools (except ISTVT) to be installed.

6.1. ACCESS Supplementary Library

There should be no changes necessary in this file unless Toolpack/1 is being installed on a smaller, non-virtual memory machine. In this case some table sizes may need to be decreased to reduce the memory requirement. Full details, if necessary, may be found in the *"TIE Installers' Guide"* [30].

This file should be split, processed by TIEMAC and then compiled.

The individual object files cannot be linked into one library. This is because different parts of the token library contain alternative routines with the same name.

If the host computer supports selective load libraries then the individual object files can be combined as follows into three libraries.

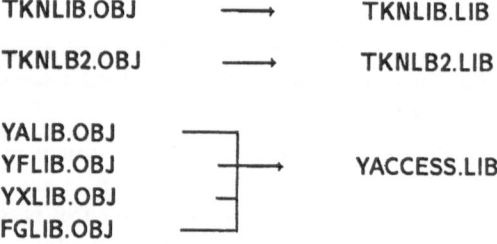

TKNLIB.OBJ   ⟶   TKNLIB.LIB

TKNLB2.OBJ   ⟶   TKNLB2.LIB

YALIB.OBJ
YFLIB.OBJ   ⟶   YACCESS.LIB
YXLIB.OBJ
FGLIB.OBJ

Other files should be left separate. If there is no selective load library provision it is recommended that the files be left separate and specified individually at the tool linking stage.

Two routines POLDUM.MAC and TKNDUM.MAC are provided to resolve external references to the polisher and lexer routines respectively. This will be necessary only on machines which do not have selective load. These routines contain dummy routines so that the actual code, which is unnecessary, need not be included in the executable module.

The selective load problem occurs when a tool calls a routine, other than ZSCAN, in TKNLIB. Although ZSCAN is not itself called some linkers/binders insist on resolving the external references in ZSCAN. TKNDUM contains dummies to resolve all these irrelevant references. Similarly POLDUM provides the same facility to the routine ZUSCAN.

6.2.   STRING Supplementary Library

STRING.FILE should be split to give the string library itself (STRING.MAC) and a test program (SSTEST.MAC). Both should be preprocessed and compiled. There should be no need to modify this library, unless it is necessary to alter the standard for Fortran variable names from the Toolpack/1 target Fortran standard. In this case the subroutine ZLEGAL, which determines the legality of variable names, should be modified to suit local requirements. The object file from STRING.MAC can be converted to a library.

The test program requires to be linked with both the STRING and TIE libraries. The test program should be run and the output checked against the example in [30].

6.3.   TABLES Supplementary Library

TABLES.FILE should be split to produce both .MAC files, that should be preprocessed and compiled, and .FTN files that should be compiled directly. The resultant object files (except from the test programs: TABTST.FTN and HLTEST.FTN) can be combined into a library. The test program should be linked with the TABLES library and run to perform a partial test of the table functions.

7.   Installing Tools

The selection of tools will depend very much on the individual user. Different sites will have different requirements. It is recommended that some basic tool functions are selected and the subset of the individual tools necessary to carry this out, is installed in the first instance. The monoliths should be installed at a later stage. Command interpreters cannot run in the stand-alone TIE regime and should be left for later.

For example, to begin with the following tools might be selected, perhaps because a priority use for Toolpack/1 is to standardise declarations.

```
ISTLX
ISTYP
ISTDS
ISTPL
ISTPO
```

Each of the tool files should be recovered from the tape, moved to a separate directory and split. This chapter will discuss in detail the installation of ISTLX extending this to more general points where appropriate.

7.1.   Installing ISTLX

The ISTLX directory should contain four files:

```
ISTLX.MAC
SCNBLK.MAC
SCNLIB.MAC
SCNLB2.MAC
```

These files all need to be preprocessed by TIEMAC and then compiled. Most tools contain include statements for files from other parts of Toolpack, often the ACCESS library. ISTLX.MAC requires the include file ZPTYPE from the ACCESS supplementary library. The *"Toolpack/1 Tool Installers' Guide"* [31] contains for each tool a list of these include files. This should be checked and a link created between the name used on the include statement and the fully qualified pathname for the file. This can be done in a number of ways, one of the easiest is to use a host operating system command, if available, to create this link. For example in Unix systems the following command could be used:

```
ln zptype /toolpack/tie/access/zptype
```

An alternative is to add TIEMAC commands to substitute the full file name. For example:

```
FILE(zptype , /toolpack/tie/access/zptype
```

The final stage in the tool installation process is that of linking or binding the various modules, from the tool file and from the supplementary libraries, to produce the executable tool.

Great care should be taken at this stage. It has been found from personal experience and from assisting Toolpack/1 installers, that the most likely stage for mistakes is here.

The *"Toolpack/1 Tool Installers' Guide"* [31] provides the installer with a list of all the files that require to be bound to install each tool. This should be carefully made into a procedure or script file specifying these files to the bind command. Particular care should be taken to ensure that the correct token library is bound. Specifying TKNLIB instead of TKNLB2 or vice versa gives no unresolved externals, but produces a tool executable that crashes.

In Section 6.1 a problem with operating systems that do not provide selective loading, was discussed. On such machines, the tools ISTLX, ISTST and some of the monoliths will require an extra dummy object file to be linked. In the case of ISTLX this file is POLDUM.

The object files of ISTLX, ISTYP, ISTDS, ISTPL, ISTSA, ISTPT and ISTPP should not be deleted as these will be required later to install the monolithic tools.

The rest of the subset of tools should be installed and tested.

The task that inspired the selection of the above subset of tools was to carry out declaration standardisation. This may be done more efficiently in terms of execution time, generation of intermediate files and tool user effort, by installing the monolithic source-to-source declaration standardiser ISTQD.

The file ISTQD.FILE should be recovered from the tape, split, and preprocessed using TIEMAC. In general monoliths require no include files from other tools or libraries. All monoliths, however, consist of a main program that calls directly or indirectly routines from other tools. Hence the binder requires the specification of, in the case of ISTQD, the routines in ISTLX, ISTYP, ISTDS and ISTPL and the usual supplementary libraries.

8.   Conclusions

The installer may now go ahead to install further tools and monoliths. The WINDOW supplementary library can be installed if the tool ISTVT is required. If the command executor is needed then the TIE embedded regime should be installed. This is outwith the scope of this chapter, but all the general principles have been dealt with in the installation of the stand-alone regime.

Tool Writing *

Wayne R. Cowell
Mathematics and Computer Science Division
Argonne National Laboratory
Argonne, Illinois, U.S.A.

"To each is given a bag of tools,
A shapeless mass and a book of rules,
And each must make, ere life is flown,
A stumbling-block or a stepping-stone."

— R. L. Sharpe,
   Stumbling Block or Stepping Stone

"I once did hold it, as our statists do,
A baseness to write fair, and labour'd much
How to forget that learning: but, sir, now
It did me yeoman's service."

— William Shakespeare,
   Hamlet, Prince of Denmark, act V, sc.2

Abstract. In addition to providing a collection of Fortran-oriented software tools, Toolpack/1 provides an environment within which new tools may be created and added to the collection. This chapter is an introduction to the features of Toolpack/1 that constitute this environment. Emphasised are the Toolpack/1 facilities that permit Fortran programs to be analyzed and manipulated in their representations as token streams and parse trees. Examples of tool-writing strategies are given, and prospective tool writers are guided to the Toolpack/1 documents that supply the detailed information they need to write tools and contribute them to the Toolpack/1 collection.

---

   * Work supported in part by the Applied Mathematical Sciences subprogramme of the Office of Energy Research, U. S. Department of Energy, under contract W-31-109-Eng-38.

A. A. Pollicini (ed.), Using Toolpack Software Tools, 251–275.

1. Introduction

The Toolpack/1 system provides an integrated collection of software tools oriented to the development and maintenance of Fortran programs. These tools are portable to any installation of Toolpack/1 and are available to the user through a command interpreter, sometimes called a user/Toolpack interface (see [17]). The Toolpack/1 system also provides an environment within which new software tools may be created and added to the collection. In this chapter we give an overview of the features of Toolpack/1 that constitute this tool-writing environment.

This is not a complete tool writer's guide; in Section 6 we will discuss the documents that contain detailed information needed by tool writers. Rather, this is a starting point for prospective writers of tools that analyze and transform Fortran. We believe that the tool-writing environment provides an opportunity to create many useful Fortran-oriented tools. We hope that this introduction to the tool-writing environment will encourage others to join the relatively small number of people on the Toolpack project who have written tools for their own use and for general distribution.

We shall not cover the writing of command interpreters – tools that provide an interface between the user and the Toolpack/1 system. However, prospective writers of command interpreters would do well to read this chapter and then consult the documents discussed in Section 6.

2. The Source Language for Toolpack/1 Tools

Toolpack/1 tools are written in Fortran. To be more precise, the tool source is "pre-Fortran" that contains symbolic or "macro" names for many predefined constants that are part of the tool-writing environment. The source must be processed by a macro expander before being compiled. For example, the macro expander would replace the lower-case words *yes* and *stderr* in the statement

```
IF (EQUAL(STR1,STR2) .EQ. yes)
+   CALL REMARK('The Two Strings Are Identical.',stderr)
```

by integer constants. These integers may be different at different Toolpack/1 installations, but their meanings are the same and the tool writer need not be concerned about their actual values. We shall interpret the above statement after a few preliminary matters are discussed, but we note at this point that the macro names are mnemonic aids to the tool writer. Tool writers often define additional macro names that pertain to their tools. These are included in files of macro definitions at the time the source is expanded.

The tool source may also contain *INCLUDE* statements that begin in column 1 and are of the form

```
INCLUDE FILNAM
```

where *FILNAM* is the name of a file whose contents are inserted into the source where the statement appears. The most common use of *INCLUDE*s is to insert the Fortran statements that define a COMMON block. The macro expander processes the *INCLUDE* statements. A macro expander called "TIEMAC" and files containing the macro definitions are part of the tool-writing environment.

3. The Tool Interface to the Toolpack/1 System

A Toolpack/1 tool manipulates data objects, defined in the tool-writing environment, using facilities defined by calls to Fortran functions and subroutines. We use the term "data object" rather than "data type" because Toolpack/1 does not provide facilities for defining data types beyond those in standard Fortran. Data objects are standard Fortran data, such as integer arrays, that have a particular structure with a special meaning. Data objects represent characters and strings as well as objects produced by the lexical analysis and parsing of Fortran programs. We will describe the data objects in the tool-writing environment, and the facilities that manipulate them, in the sections that follow. The Toolpack/1 functions and subroutines that are invoked by a tool are called *primitives*. Services provided by the primitives include input and output of data from and to files, including special files created by the Toolpack/1 system. Tool writers should refer to [17] for a discussion of the file systems in the Toolpack/1 operating environment.

A new tool may be incorporated into the Toolpack/1 tool collection if it (1) uses predefined macro names for all Toolpack-dependent and installation-dependent constants, (2) uses Toolpack/1 primitives for all input and output, and (3) conforms to certain conventions about initiation and termination, to be discussed in Section 5.2. Such a tool is completely portable to any Toolpack/1 installation, because the primitives provide exactly the same services at every installation, no matter how they are implemented on the host machine. Moreover, the tool may be invoked by any user/Toolpack interface that conforms to Toolpack/1 conventions.

4. Characters and Strings

Characters and *strings* are basic data objects. The tool-writing environment represents characters as small predefined integers and provides for the ASCII character set. If the full ASCII character set is not available in a given host system, the implementation of the primitives must provide for a suitable mapping to the host character set.

A string in the tool-writing environment is called an *IST string* and is a sequence of elements, representing characters, in an integer array. (The elements are called "characters" although, of course, their Fortran data type is *integer* .) The IST string begins with the first element of the array and is terminated by a special integer called the *end-of-string* character. The dimension of the array must be large enough to accommodate the IST string. Elements of the array following the end-of-string character have no meaning.

Many primitives take IST strings as arguments. The example statement in Section 2 calls primitives **EQUAL** and **REMARK** to achieve the result that if IST strings STR1 and STR2 are equal character by character, then "The Two Strings Are Identical" is output to the standard error file.

Characters have macro names in the tool-writing environment to facilitate the specification of constant characters and IST strings. For example, the specification statements

```
INTEGER TOK1(9)
DATA TOK1/bigt,leto,letk,lete,letn,lets,period,dig1,eos/
```

would define an IST string containing "Tokens.1" named *TOK1* . Note that the string ends with the end-of-string character whose macro name is *eos* .

To exhibit several primitives that take IST strings as input, we present the following example. The tool writer wishes to open a sequential file named "Tokens.1" for reading, and to place an error message in the standard error file if the open fails (for example, if a file by that name does not exist). He writes

```
FD = OPEN(TOK1,read)
IF (FD .EQ. err) THEN
   CALL PUTLIN(ERRMSG,stderr)
   CALL ZPTMES(TOK1,stderr)
END IF
```

where ERRMSG is an IST string, say, "Could Not Open File:". The primitive **OPEN** returns either a file descriptor, that is, an integer used to reference the file in input/output primitives, or the integer *err* if the open fails. **PUTLIN** writes a string to the sequential file specified by its second argument, which is a file descriptor. In the above example, the preset file descriptor *stderr* designates the standard error file. Similarly, **ZPTMES** writes an IST string to a file but follows it with the *newline* character that signals the end of a line in the file. Hence, if the open fails, the error message

```
Could Not Open File:  Tokens.1
```

would appear on the standard error file. If the open is successful, *FD* may be used as a file descriptor in calls to primitives that read data from *Tokens.1* .

5.   Tokens, Nodes, and Symbol Attributes

Tools that analyze and transform a Fortran program are based on the ability to examine and manipulate suitable representations of the program. The Toolpack/1 tool collection includes two tools – the lexer (ISTLX), also called the "scanner", and the parser (ISTYP) – that create representations of a program called, respectively, the *token/comment stream* and the *parse tree/symbol table*. Components of these representations, namely, *tokens* (in the token/comment stream), *nodes* (of the parse tree), and *symbol attributes* (from the symbol table) are data objects in the Toolpack/1 tool-writing environment.

5.1.   Accessing the Token/Comment Stream

Tokens are "atomic" units of Fortran that cannot be subdivided into any lower level Fortran objects. For example, "(", "+", and "CONTINUE" are tokens, as is any name in the program. To illustrate further, the assignment statement

        20   R(J) = R(J) - P(J,5)*Q(5)

may be represented as the sequence of tokens shown in Fig. 1. Some token types (e.g., <name> and <integer constant>) have strings associated with them; in the figure, the associated string is shown under the token. A special end-of-statement token marks the end of the statement. The Toolpack/1 scanner outputs a file containing the sequence of tokens together with their associated strings, except for strings associated with tokens of type <comment>. These are stored in a second file and indexed to their place in the token sequence. The two files together constitute the *token/comment stream*.

```
<integer constant> <name> <left parenthesis> <name>
        20              R                        J
<right parenthesis> <equals> <name> <left parenthesis>
                              R
<name> <right parenthesis> <minus> <name> <left parenthesis>
  J                                  P
<name> <comma> <integer constant> <right parenthesis>
  J                     5
<star> <name> <left parenthesis> <integer constant>
        Q                                5
<right parenthesis> <end of statement>.
```

Figure 1. The Token Stream Representation of
20   R(J) = R(J) − P(J,5)*Q(5)

Using the file descriptors of these files, a token channel is created by **ZTKGTI** (for input) or **ZTKPTI** (for output). These primitives return channel descriptors used by the primitives that read or write tokens.

The tool writer need not be concerned about the internal representations of tokens or the operation of token channels. He can read and write tokens and determine their types using primitives and predefined names for the token types. We describe one such primitive to give the flavour of such operations. Each successive execution of

```
CALL ZGETTK(TYPE,LENGTH,STRING,INCHAN,STATUS)
```

reads a token from the token/comment stream whose descriptor is *INCHAN*. *TYPE* is an integer whose value specifies the type of the token. Symbolic names for the token types are set by parameter statements in the file *ZPTYPE* that is included in any program unit in which **ZGETTK** is called. The tool writer refers to token types by these symbolic names; for example, *TEQUAL*, *TNAME*, *TPLUS*, and *TZEOS* are the predefined names in a Fortran statement for token types <equal>, <name>, <plus>, and <end of statement>, respectively.

The argument *LENGTH* in the above call is the length of the string, if any, associated with the token. If *LENGTH* is greater than 0, there is an associated string and it is returned as the IST string *STRING*. *STATUS* indicates the return status of the call. Its return value is *noerr*, (successful return), *err*, (error), or *eof* (end-of-file).

5.2.   A Declaration Inserter Tool

Given data objects and primitives, it is reasonable to ask how they are assembled into a software tool. Prospective tool writers are advised to study existing tools as models. In that spirit we will now sketch the design and operation of a tool that uses the concepts we have discussed so far and a few others that will be brought out in the description, in particular, the conventions for initiation and termination that enable a tool to operate within Toolpack/1. This is a relatively simple tool by comparison with many of the tools in Toolpack/1, but it is non-trivial enough for a real example.

The tool is a *declaration inserter* that enables the programmer to specify the data types of variables and functions by writing directives in the form of comments in the code at the point where he is working. Programmers often wish to override the type indicated by the first letter of a variable or to call a function whose name does not correspond to the type of the data returned. In either case they must remember to explicitly type the name; forgetting to do this is a common source of errors. The tool is named ISTDI and its source is contained in the Appendix. The main program, named *ISTDI*, is listed first and the remaining program units are listed in alphabetical order of their names.

Toolpack/1 has primitives that decode special comments, called *source-embedded directives*, of the form "*xxabc..." where the two characters *xx* are the *identification* and *abc...* is any string, called the *body,* not longer than the remainder of the 80-character line. The source-embedded directives processed by the declaration inserter tool are of the form "*DIyy name" where *yy* is a two-letter code indicating the data type (*re* for REAL, *in* for INTEGER, etc.) of *name* . The programmer inserts these directives in his code anywhere in the program unit, presumably near the line where *name* is first used. Each name whose type is declared requires a separate source-embedded directive.

ISTDI reads the token/comment stream, detects the source-embedded directives with identification *DI* (which is case-insensitive) and stores the associated bodies in a table. The table contains a separate list for each program unit. The tool then creates a replacement token/comment stream, in which tokens that compose the declarations specified in the table are inserted into the declaration part of the program unit and the source-embedded directives with identification *DI* are deleted.

ISTDI takes full advantage of the integration of the tool set in its operation. The lexer (ISTLX) is invoked first to produce the token/comment stream. The declaration inserter then produces a replacement token/comment stream. The formatter (ISTPL) could then be invoked to recover Fortran text from the replacement token/comment stream. However, a better approach is to invoke the parser (ISTYP) on the replacement stream to create a parse tree/symbol table, then the declaration standardiser (ISTDS) whose output is a new token/comment stream, and then ISTPL on the new stream. The resulting Fortran text not only will have the declarations inserted but will have a rebuilt, comment-annotated declarative part. These questions of usage should be covered in the documentation of the tool. In particular, the people who build command interpreters need to know; they, in turn, will create a user interface that enables users to make optimal use of the tool.

The first executable statement in the main program of any tool must be

```
CALL ZINIT
```

This call opens and/or creates various special files and establishes contact with the Toolpack/1 file systems (see [17]). The tool then uses the primitive **GETARG** to read input parameters from an inter-process file set up by the command interpreter. A typical call is

```
IF (GETARG(1,TKNPTH,maxpath) .EQ. eof) CALL NAMES(1,TKNPTH)
```

which reads the first argument from the inter-process file (in this case, the name of the input token stream) into the IST string *TKNPTH*, truncating it to *maxpath* characters if necessary (*maxpath* is the macro name for the longest possible string representing a file name; it is a system-dependent parameter set by the installer). If GETARG fails (for example, if there is no inter-process file), then the prompting subroutine, *NAMES* (see Appendix), is called to obtain the name of this parameter from the user. This method of obtaining input parameters is independent of how a command interpreter sets up the inter-process file and so is portable from one Toolpack/1 installation to another.

ISTDI has four input arguments obtained in this way: the names of the input token file, input comment file, output (replacement) token file, and output (replacement) comment file. Having obtained the names, the tool opens the two input files that were created by ISTLX and creates the two output files. A typical file opening sequence is

```
IODTKN=OPEN(TKNPTH,read)
IF (IODTKN.EQ.err) CALL ERROR('Can''t Open Token/In Path.')
```

(The primitive **ERROR** transmits a message to the standard error file and terminates the operation of the tool.) From the file descriptors obtained in this way, channel descriptors for the input and output token streams are obtained as shown in the channel initialisation calls to **ZTKGTI** and **ZTKPTI** in the main program.

The tool is now ready for its first pass through the token/comment stream, in which it records the insertion directives. **ZGETTK** is repeatedly called and the token type examined. The token *TEND* signals the end of a program unit and the token *TZEOF* the end of the file. The comments are tokens of type *TCMMNT*. Each comment is tested to see if it is a source-embedded directive with identification *DI*. This is accomplished by the function *CHKINS* where the primitive **ZSEDID** tests whether a string has the form of a source-embedded directive. The call is

```
STATUS = ZSEDID(STRING,BIND,ID,BODY)
```

in which *STATUS* is returned as an integer whose macro name is *yes* or *no,* depending on whether the IST string *STRING* is or is not a source-embedded directive. If it is, the identification is returned as the IST string *ID* and the body as IST string *BODY.* (*BIND* is an integer that may be ignored – it is a remnant of an earlier version of the routine and has no current meaning.) Each source-embedded directive with identification *DI* is recorded in a table; the bookkeeping details are straightforward and need not concern us at this point.

When the end of the token/comment stream is reached, the tool returns to the beginning of the stream with the code shown immediately following

```
30 CONTINUE
```

in the main program *ISTDI.*

On the second pass through the token/comment stream, each token is read and written to the output token/comment stream unless it is a source-embedded directive with identification *DI,* in which case it is not written to the output stream, or unless a point is reached where there are declaration statements to be inserted, in which case the source-embedded directives in the table for that program unit are decoded and tokens for the insertions are written to the output stream. See the comments and code immediately following

```
40 CONTINUE
```

in the main program *ISTDI*. Most of the work is done in subroutines *STEPDI* and *DECODE*. Token writing is accomplished by the primitive **ZPUTTK** whose arguments are the token's type, length of associated string (or 0), associated string (if it has one) and the channel descriptor of the output channel.

The algorithm employed by *STEPDI* to determine where declarations may be inserted into a program unit is of interest. Assume that we start reading tokens at the beginning of the program unit. Pass comments (except the insertion directives) through to the output token/comment stream. Examine the first pair of tokens of the first statement to determine whether the statement is a program unit statement (a PROGRAM or SUBROUTINE or BLOCK DATA statement or a function subprogram statement), or an IMPLICIT statement. (Two tokens must be examined because of the possibility of a function subprogram statement that declares the type of the function.) If the statement is any of these, pass the tokens for the statement through to the output token/comment stream and look at the first two tokens of the next statement. If the statement is none of these, then, since we started the search at the beginning of the program, the statement is such that a type declaration may immediately precede it.

At the end of the second pass the tool is finished. It must terminate with the call

```
CALL ZQUIT(STATUS)
```

where *STATUS* is an integer whose macro name is one of *ok, warning, err,* or *fatal,* depending on the termination condition of the tool. **ZQUIT** will close all opened files, inform the command interpreter of the termination condition, and pass control back to the command interpreter. The command interpreter can then schedule further tools such as the sequence described above. A call to **ZQUIT** or **ERROR** (which, in turn, calls **ZQUIT**) is the only legal way for a tool to terminate.

5.3.   Accessing the Parse Tree/Symbol Table

The parser, ISTYP, takes a token/comment stream as input and, by referring to a Fortran grammar, produces a representation of the program as a *parse tree,* together with an associated *symbol table.* The output from the parser consists of three files: the tree file, the symbol table file, and the comment index file. Conceptually, the output may be thought of as a tree composed of nodes and branches. For example, the assignment statement whose token/comment stream is shown in Fig. 1 is represented as the subtree shown in Fig. 2. (We shall draw our trees growing downward with the root at the top of the figure.) The types of the nodes are shown in brackets. Nodes of certain types, such as [name] and [integer constant], have associated symbols in the symbol table. These are shown below the node.

Node types are represented as integers whose macro names are defined in the file *YNODES*. An *INCLUDE YNODES* statement in the source enables tool writers to use them.

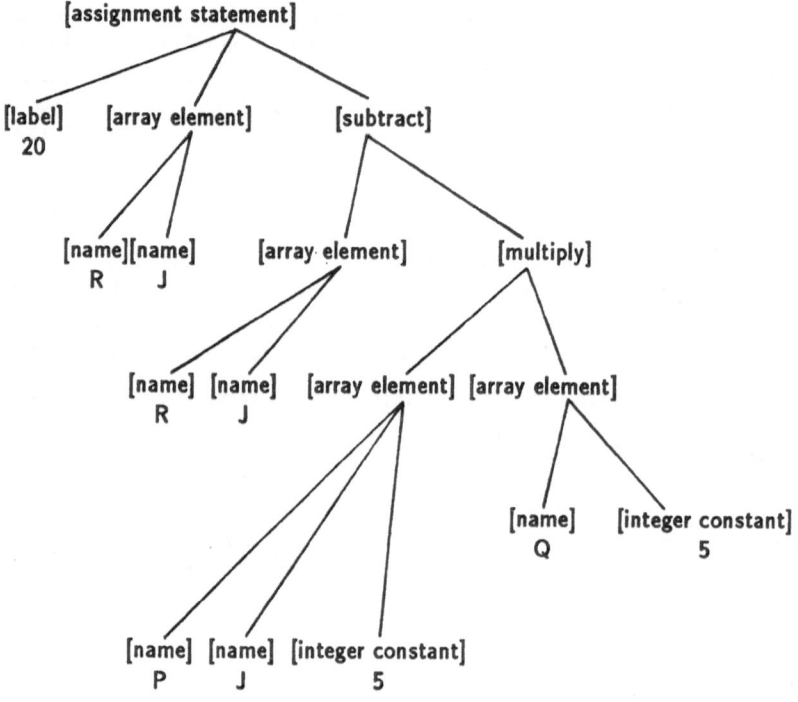

Figure 2. The Parse Tree Representation of
20   R(J) = R(J) − P(J,5)*Q(5)

Nodes are represented by positive integers. The primitive **ZYNTYP** takes a node as input and returns its type. The primitives **ZYDOWN, ZYNEXT, ZYPREV,** and **ZYUP** permit the tool to "walk about" on the parse tree. To illustrate this, we refer to Fig. 2 and consider the nodes of type [assignment statement], [label], [array element], and [subtract] at the top of the subtree. The macro names for these types are *N_ASGN, N_LABEL, N_ARELM,* and *N_MINUS* respectively. Let these nodes be represented by N1, N2, N3, and N4 in the same order. Then N2 = ZYDOWN(N1),

N3 = ZYNEXT(N2), N4 = ZYNEXT(N3), N3 = ZYPREV(N4), N2 = ZYPREV(N3), and N1 = ZYUP(N2) = ZYUP(N3) = ZYUP(N4). ZYNEXT(N4) returns 0 which signals that there is no next node.

Nodes with no downward branches are called *leaves*. Nodes with associated symbols in the symbol table are always leaves and are called *named leaves*. ZYDOWN(N2) returns a negative integer which signals that N2 is a named leaf. Using primitives to be discussed below, the returned integer can be used to access the symbol table to determine attributes of the symbol associated with N2.

The root of the parse tree (type *N_ROOT*) has branches to nodes for the program units (type *N_PROGRAM*) which, in turn, have branches to the statements. To design his code, the tool writer needs to know this general structure and the structure of various kinds of statements. For example, a node of type *N_ASGN* always has either two or three branches. If it is labeled, as in Fig. 2, there are three; the first branch is to a node of type *N_LABEL*, the second to a subtree representing the left side of the assignment, and the third to a subtree representing the right side of the assignment. If it does not have a label, there are only the branches to the left and right sides. The parse tree viewing tool, ISTVT, is an aid to tool writers in determining the structure of Fortran statements.

To illustrate tree walking, let us suppose a tool is examining the statements in a program to determine some property of interest. It would start at the root, execute **ZYDOWN** to reach the first program unit, execute another **ZYDOWN** to reach the first statement of the first program unit, and then proceed from any one statement to the next with **ZYNEXT** until 0 is returned. To proceed to the next program unit, the tool would execute **ZYNEXT** from the current program unit node. If *NODE* is the node for a statement, the following sample code sets the integer variable *VALUE* to *yes* if the statement is an assignment statement in which the left side is an array element, and to *no* otherwise.

```
VALUE = no
IF (ZYNTYP(NODE) .EQ. N_ASGN) THEN
    PTR = ZYDOWN(NODE)
    IF (ZYNTYP(PTR) .EQ. N_LABEL) PTR = ZYNEXT(PTR)
    IF (ZYNTYP(PTR) .EQ. N_ARELM) VALUE = yes
END IF
```

The Appendix to the Chapter "DO Loop Transforming Tools" contains the source of a routine that illustrates tree walking.

It is frequently necessary to systematically traverse all the nodes of a subtree. This situation would arise, for example, in comparing the expressions rooted at two different nodes or in displaying the structure of the subtree representing a statement. The approach is as follows. Build a stack (last in, first out) of nodes. Proceed downward (**ZYDOWN**) as long as possible, pushing nodes onto the stack. When down is no longer possible (a leaf node) try to go to the next node (**ZYNEXT**). If that is possible, try to go down from that node. When neither down nor next is possible, pop the stack until next is possible, then try down again. Continue in this way until returning to the root of the subtree.

The symbols in a Fortran program are the names of COMMON blocks, program and subprogram units, variables, parameters, statement functions, entry points, and labels. When the parser builds the parse tree from the token/comment stream, it records the attributes of symbols in a symbol table. For example, when the statement

```
INTEGER X
```

is scanned by the lexer ISTLX, a token representing *INTEGER* and a token representing X are placed in the token stream. The connection between these tokens is recognised by ISTYP, which identifies X as an explicitly typed integer name and records this information in the symbol table. If the statement

```
X = 1
```

marks the first appearance of X in an executable statement, ISTYP adds to the list of attributes the information that X is a variable (the INTEGER type declaration did not ensure that) and appears on the left-hand side of an assignment statement. Then if the statement

```
X = ABS(X - 5)
```

appears, ISTYP adds to the symbol table the information that X appears in an expression. ISTYP also adds the information that *ABS* is a standard Fortran intrinsic function called by the program and appearing in an expression. The name of a symbol may be regarded as one of its attributes.

The tool-writing environment provides various primitives for accessing the information in the symbol table. Tools such as ISTVS, the symbol table viewer, use these primitives. The use of macro names greatly simplifies the tool writer's task of coping with detailed, tightly encoded information.

An important instance of accessing the symbol table occurs when a walk of the parse tree encounters a named leaf and information about the associated symbol, such as its name, is required. If the node is, say, *NODE*, then, as mentioned above, *ZYDOWN(NODE)* returns a negative number, say, POINTR. The primitive **ZYGTSY** uses -POINTR to obtain the entries in the symbol table associated with *NODE*. The call

```
CALL ZYGTSY(-POINTR, VALUES)
```

fills an integer array *VALUES*, whose dimension has the macro name *symbol_size*, with symbol table entries for the symbol associated with the node. In the current version of Toolpack/1, *symbol_size* is 8; that is, there are at most eight integers associated with each symbol. These must be decoded to determine various attributes. This encoded information varies with the symbol type; for example, labels and variables both have names, but variables have data types while labels do not.

Decoding the entries in *VALUES* is facilitated by the fact that these entries have macro names and may be referenced mnemonically using macro names for the index. For example, the following statements set *REALV* to *yes* if the symbol is a real variable and to *no* otherwise.

```
      REALV = no
      IF (VALUES(symbol_type) .EQ. S_VAR .AND.
     +      VALUES(name_dtype) .EQ. type_real) REALV = yes
```

As a further illustration, suppose we have determined that a symbol is a real variable and wish to determine whether it is explicitly typed. *VALUES(name_status)* is an integer in which "1" in any of the 17 least significant bit positions signifies that the symbol possesses the attribute associated with that bit and "0" signifies that the symbol does not possess that attribute. Powers of 2 from 1 to 65536 have macro names to facilitate masking operations to decode *VALUES(name_status)*. In particular, *explicit_typ* is the macro name for the power of two corresponding to the bit position in *VALUES(name_status)* that is set to "1" when the symbol is explicitly typed. The following code writes a message to the standard output file stating whether the variable is explicitly or implicitly typed.

```
C EXTYP is a mask with 1 in the "explicitly typed" position.
C ZIAND is a primitive that returns the bitwise logical AND
C        of its arguments.
C ZMESS is a primitive that writes a Fortran 77 string to a file.
      EXTYP = explicit_typ
      IF (ZIAND(EXTYP,VALUES(name_status) .EQ. EXTYP) THEN
         CALL ZMESS('Explicitly Typed.',stdout)
      ELSE
         CALL ZMESS('Implicitly Typed.',stdout)
      END IF
```

VALUES(symbol_name) is a pointer to a string table containing the name of the symbol. The name can be recovered by using the primitive **ZYGTST** whose call is

```
      CALL ZYGTST(VALUES(symbol_name),STRING)
```

where the output IST string *STRING* contains the name.

6. Documentation for the Tool Writer

The Toolpack/1 tool writer should, first of all, be a Toolpack/1 user so that he can make use of Toolpack/1 tools in the construction of new tools. For example, tools like ISTAL and ISTVS are useful in locating untyped variables and functions and anomalous uses of variables that could be errors. Moreover, as a Toolpack/1 user, the tool writer will appreciate what other tools do, what sort of information is helpful to users, etc. As a user, he will be familiar with the user/Toolpack interface at his installation and will be prepared to think of his tool as a member of an integrated collection. Hence, he should be familiar with the local user documentation and the *"Toolpack/1 Introductory Guide"* [17].

The primitives introduced in this chapter, and many others in the Toolpack/1 tool-writing environment, are documented in five Toolpack/1 documents ([12], [34], [35], [38] and [39]) that correspond to five collections. The primitives described in [35] constitute the Toolpack/1 Virtual Machine that serves as the tool/host interface. Reference [35] also includes a list of the macro names for characters, pre-connected files, and other basic quantities, as well as a description of the Toolpack/1 file systems.

It is recommended that tool writers think of the file system as "flat". This means that they can ignore any primitives that deal with directory handling. These may very well be used by command interpreters but, as we indicated earlier, we are not covering that subject. Questions about the Toolpack/1 operating regime, tool invocation, and implementation of the primitives are left to installers, command interpreter writers, and system developers. It is enough for tool writers to follow the simple conventions of the example tool in Section 5.2. This segregation of issues is an important feature of the Toolpack/1 tool-writing environment. Further, we recommend that tools create only sequential files with alphanumeric names.

The primitives documented in [35] are divided into three categories, corresponding to three "libraries": flow of control (e.g., **ZINIT** and **GETARG**), input/output (e.g., **REMARK**, **PUTLIN**, and **ZPTMES**), and the so-called common library (e.g., **EQUAL** and **ZIAND**). There are two approaches to the implementation of the flow-of-control library and the input/output library. The choice of implementation at load time determines the operating regime for Toolpack/1 but, as noted above, this is not an issue for tool writers who should regard the primitives as "black boxes" that function according to their descriptions.

References [12], [34], [38] and [39] describe the so-called supplementary libraries. Reference [38] describes primitives in the ACCESS library that provide access to the token/comment stream and parse tree/symbol table. These are illustrated in this chapter by **ZGETTK**, **ZPUTTK**, **ZYDOWN**, and **ZYGTSY**. The ACCESS library also includes the parse tree "flattening" primitives that map subtrees of the parse tree to sequences of tokens. These are not illustrated in this chapter but are covered in the Chapters "Toolpack/1 Fortran Transformers" and "DO Loop Transforming Tools". Tree-flattening primitives are used by Toolpack/1 transformation tools such as the precision transformer, the declaration standardiser, and the tools that transform DO loops to improve performance on vector machines. These tools extract information from the parse tree and the symbol table and then write a new token/comment stream representing the transformed program.

Reference [34] describes the STRING library, a set of primitives that manipulate strings; ZSEDID is an example. Reference [39] describes the TABLES library that contains primitives that set up and access data structures like linked lists, tables, stacks, and queues. The documentation generation aid ISTAL and the parser ISTYP are examples of Toolpack/1 tools that use these primitives. Reference [12] describes the WINDOW library, a set of primitives for performing terminal screen manipulation in a portable manner. The parse tree viewing tool, ISTVT, uses primitives from this collection.

7.   Contributing Tools to Toolpack/1

As the number of Toolpack/1 installations grows, so also do opportunities to share software tools written for integration into the Toolpack/1 environment. One can draw an analogy with the sharing of mathematical software, which requires the installation of a Fortran compiler – certainly an easy requirement to meet in the present computer milieu! One can imagine libraries of tools, beginning with the current collection.

In addition to their fundamental technical contributions to the Toolpack project, the Numerical Algorithms Group (NAG) has assumed a leadership rôle in distributing Toolpack/1. This rôle includes the maintenance of information about Toolpack-related activities and the coordination of efforts to strengthen Toolpack/1 in various ways, including the addition of tools to the distributed collection. The contribution of tools to Toolpack/1 is discussed in [32]. An important point is that any tool writer who is contemplating a major effort to develop a new tool can obtain information from NAG about related efforts and about the requirements to be met by a tool that is contributed for inclusion in Toolpack/1. It is recommended that the writer of a tool destined for contribution to Toolpack/1 contact NAG at an early stage of his project.

Whether or not he expects to contribute his tool, the prospective tool writer should read [32] for valuable information to supplement what is covered in this chapter. In particular, [32] discusses the preparation of supplementary libraries, exhibits two example tools, gives some useful hints about minimising non-portable constructions, and mentions some design aspects of Toolpack/1 that are in transition.

Acknowledgements.   The author thanks Burton Garbow and Gail Pieper for their comments on the manuscript.

Appendix:
Source for the tool ISTDI

```
C     *** I S T D I ***

      PROGRAM ISTDI

C Process a token stream derived from a Fortran source file containing
C source-embedded directives that specify the declaration of names in
C the program.  The syntax of a directive is a comment of the form
C
C  *$DI$  xx  $<$remainder of declaration$>$
C
C where xx is one of IN (for INTEGER), RE (for REAL), DB (for DOUBLE
C PRECISION), CO (for COMPLEX), LO (for LOGICAL), or CO (for COMPLEX).
C
C Record the declarations on the first pass through the token stream.
C On the second pass, insert corresponding declarations and remove the
C directives.
C
C The file ZPTYPE is in the access library.  It contains a set of
C parameter definitions that are the integer codes for the token types.
INCLUDE ZPTYPE

      INTEGER TKNPTH(maxpath),CMTPTH(maxpath),
     +        TKNOUT(maxpath),CMTOUT(maxpath),
     +        STRING(maxtoklen)

      INTEGER TOKTYP,LENGTH,STATUS,IODTKN,IODCMT,INCHN,
     +        IODTKO,IODCMO,OUTCHN,PUCNT,COUNT

      INTEGER BODY(81),TYPE(2),LEN(2),STR(maxtoklen,2),
     +        BEG(50),FIN(50),INSTAB(75,500),DUMMY(2)

      INTEGER GETARG,OPEN,CREATE,CHKINS,ZTKGTI,ZTKPTI

      EXTERNAL ZINIT,GETARG,OPEN,SCOPY,ERROR,NAMES,
     +         CREATE,SEEK,ZQUIT,ZMESS,CHKINS,ZGETTK,
     +         STEPDI,ZPUTTK,DECODE,ZTKGTI,ZTKPTI,ZTKGTQ

      DATA DUMMY(1)/eos/

C ZINIT initialises the link to TIE.
```

```
      CALL ZINIT

C Read paths from the inter-process file.  If it does not exist
C or if an entry is not present, call NAMES to prompt user.
      IF (GETARG(1,TKNPTH,maxpath).EQ.eof) CALL NAMES(1,TKNPTH)
      IF (GETARG(2,CMTPTH,maxpath).EQ.eof) CALL NAMES(2,CMTPTH)
      IF (GETARG(3,TKNOUT,maxpath).EQ.eof) CALL NAMES(3,TKNOUT)
      IF (GETARG(4,CMTOUT,maxpath).EQ.eof) CALL NAMES(4,CMTOUT)

C Open/Create required files.
      IODTKN=OPEN(TKNPTH,read)
      IF (IODTKN.EQ.err) CALL ERROR('Can''t Open Token/In Path.')
      IODCMT=OPEN(CMTPTH,read)
      IF (IODCMT.EQ.err) CALL ERROR('Can''t Open Comment/In Path.')
      IODTKO=CREATE(TKNOUT,write)
      IF (IODTKO.EQ.err) CALL ERROR('Can''t Create Token/Out Path.')
      IODCMO=CREATE(CMTOUT,write)
      IF (IODCMO.EQ.err) CALL ERROR('Can''t Create Token/Out Path.')

C Initialise the input and output channels for the original and
C modified token streams, respectively.
      INCHN = ZTKGTI(1,IODTKN,IODCMT)
      IF (INCHN .LE. 0) CALL ERROR('Can''t Initialise Input Channel.')
      OUTCHN = ZTKPTI(1,IODTKO,IODCMO)
      IF (OUTCHN .LE. 0) CALL ERROR('Can''t Initialise Output '
     +                                       //'Channel.')

C First pass through token stream.  Construct table of declarations.

C PUCNT is the program unit count.  BEG(PUCNT) and FIN(PUCNT) are,
C respectively, the beginning and end indices in INSTAB of the
C insertions for that program unit.

      PUCNT = 1
      BEG(1) = 1
      FIN(1) = 0
   20 CONTINUE

C Read a token.
      CALL ZGETTK(TOKTYP,LENGTH,STRING,INCHN,STATUS)
      IF(STATUS.EQ.err.OR.STATUS.EQ.eof) CALL ERROR(
     +'Error In Reading Token Stream - First Pass.')

C If we have reached the end of the token stream, go on to the
C second pass.

      IF (TOKTYP.EQ.TZEOF) GO TO 30

C If we have reached the end of a program unit, update BEG and PUCNT.
```

```
C FIN is updated as new insertions are added to INSTAB.

      IF (TOKTYP.EQ.TEND) THEN
         BEG(PUCNT+1) = FIN(PUCNT) + 1
         FIN(PUCNT+1) = FIN(PUCNT)
         PUCNT = PUCNT + 1
         GO TO 20
      ENDIF

C If the token is a comment, check whether it is a source-embedded
C directive with identification "DI" (case insensitive).  If so, store
C the body of the directive in the insertion table.

      IF (TOKTYP.EQ.TCMMNT) THEN
         IF (CHKINS(STRING,BODY) .EQ. yes) THEN
            FIN(PUCNT) = FIN(PUCNT) + 1
            CALL SCOPY(BODY,1,INSTAB(1,FIN(PUCNT)),1)
         END IF
      END IF
      GO TO 20

C Second pass.  Decode the source-embedded directives and transfer the
C token stream files to new files with appropriate type declarations
C inserted.

C Return to beginning of token stream and reinitialise the input
C channel.

   30 CONTINUE
      CALL SEEK(0,IODTKN)
      CALL SEEK(0,IODCMT)
      CALL ZTKGTQ(INCHN)
      INCHN = ZTKGTI(1, IODTKN, IODCMT)
      IF (INCHN .LE. 0) CALL ERROR('Can''t Reopen Input Channel.')

C For each program unit, transfer the tokens, removing the original
C insertion directives, until reaching the point where insertions
C are to be made.  Then decode the instructions and make the
C insertions.  COUNT is the number of the program unit.

      COUNT = 1
   40 CONTINUE

C Transfer tokens, removing original directives, up to the place in
C the code where insertions are made.  Return the first two tokens,
C specified by type, length, and associated strings, that are to
C follow the insertion.
      CALL STEPDI(INCHN,OUTCHN,STATUS,TYPE,LEN,STR)
      IF (STATUS .EQ. eof) THEN
```

```
C We have processed all the program units and reached the end-of-file
C token.  Output it and terminate.
        CALL ZPUTTK(TZEOF,O,DUMMY,OUTCHN)
        CALL ZMESS('[ISTDI Normal Termination].', stderr)
        CALL ZQUIT(ok)
        STOP
     END IF

C If program unit COUNT has insertions, decode the instructions in
C INSTAB and make the insertions.
     IF (FIN(COUNT) .GE. BEG(COUNT))
    +      CALL DECODE(INSTAB,BEG(COUNT),FIN(COUNT),OUTCHN)

C Output the remaining tokens for program unit COUNT, removing
C original directives.  The first two to be output are returned
C by STEPDI; the remainder are transferred from input to output
C token/comment files.

C Output the first two tokens.
     IF (TYPE(1) .NE. TCMMNT .OR. CHKINS(STR(1,1),BODY) .EQ. no)
    +      CALL ZPUTTK(TYPE(1),LEN(1),STR(1,1),OUTCHN)
     IF (TYPE(2) .NE. TCMMNT .OR. CHKINS(STR(1,2),BODY) .EQ. no)
    +      CALL ZPUTTK(TYPE(2),LEN(2),STR(1,2),OUTCHN)

C Output the remaining tokens.
  50 CONTINUE
     CALL ZGETTK(TOKTYP,LENGTH,STRING,INCHN,STATUS)
     IF(STATUS.EQ.err.OR.STATUS.EQ.eof) CALL ERROR(
    +'Error In Reading Token Stream - Second Pass.')
     IF (TOKTYP .NE. TCMMNT .OR. CHKINS(STRING,BODY) .EQ. no)
    +      CALL ZPUTTK(TOKTYP,LENGTH,STRING,OUTCHN)

     IF (TOKTYP .EQ. TEND) THEN
C We have reached the end of the program unit.  Output the
C end-of-statement token for the END statement and proceed
C to the next program unit.
        CALL ZGETTK(TOKTYP,LENGTH,STRING,INCHN,STATUS)
        IF(STATUS.EQ.err.OR.STATUS.EQ.eof) CALL ERROR(
    +      'Error In Reading Token Stream - Second Pass.')

        CALL ZPUTTK(TOKTYP,LENGTH,STRING,OUTCHN)
        COUNT = COUNT + 1
        GO TO 40
     ELSE
        GO TO 50
     END IF

     END
```

```
C     *** C H K I N S ***

      INTEGER FUNCTION CHKINS(STRING,BODY)

C Check whether STRING is a source-embedded directive with
C identification "DI".  If so the value of the function is
C 'yes' and the body of the SED is returned in BODY.  If not,
C the value of the function is 'no'.

      INTEGER STRING(maxtoklen),BODY(81)
      INTEGER ID(3),STATUS,BIND,DISTR(3)

      INTEGER EQUAL,ZSEDID

      EXTERNAL ZSEDID,EQUAL

      DATA DISTR/letd,leti,eos/

      STATUS = ZSEDID(STRING,BIND,ID,BODY)
      IF (STATUS.EQ.yes) THEN
         IF (EQUAL(ID,DISTR).EQ.yes) THEN
            CHKINS = yes
            RETURN
         ELSE
            CHKINS = no
            RETURN
         END IF
      ELSE
         CHKINS = no
         RETURN
      END IF

      END


C     *** D E C O D E ***

      SUBROUTINE DECODE(INSTAB,START,STOP,TCHANO)

C Decode the insertion instructions from INSTAB and output the
C corresponding specification statements to the token output channel
C TCHANO from the instruction with index START to the instruction
C with index STOP.

INCLUDE ZPTYPE

      INTEGER INSTAB(75,500),START,STOP,TCHANO

      INTEGER COUNT,SPEC(3),NAME(71),POINT,DUMMY(2)

      INTEGER EQUAL,LENGTH
```

```
      EXTERNAL EQUAL,ZPUTTK,LENGTH,SKIPBL,ZTOCAP,SCOPY,REMARK,ZPTMES

      INTEGER IN(3),RE(3),DP(3),LO(3),CO(3),CH(3)

      SAVE

      DATA IN/bigi,bign,eos/
      DATA RE/bigr,bige,eos/
      DATA DP/bigd,bigp,eos/
      DATA LO/bigl,bigo,eos/
      DATA CO/bigc,bigo,eos/
      DATA CH/bigc,bigh,eos/

      DATA DUMMY(1)/eos/

      COUNT = START

   10 CONTINUE
C Copy the specification part of the instruction into SPEC and change
C to all caps for comparison.

      POINT = 1
      CALL SKIPBL(INSTAB(1,COUNT),POINT)
      SPEC(1) = INSTAB(POINT,COUNT)
      SPEC(2) = INSTAB(POINT+1,COUNT)
      SPEC(3) = eos
      CALL ZTOCAP(SPEC)

C Copy the name part of the instruction into NAME.

      POINT = POINT + 2
      CALL SKIPBL(INSTAB(1,COUNT),POINT)
      CALL SCOPY(INSTAB(1,COUNT),POINT,NAME,1)

C Decode the instruction and output the appropriate sequence of tokens.

      IF (EQUAL(SPEC,IN) .EQ. yes) THEN
         CALL ZPUTTK(TINTEG,0,DUMMY,TCHANO)
         CALL ZPUTTK(TNAME,LENGTH(NAME),NAME,TCHANO)
         CALL ZPUTTK(TZEOS,0,DUMMY,TCHANO)

      ELSE IF (EQUAL(SPEC,RE) .EQ. yes) THEN
         CALL ZPUTTK(TREAL,0,DUMMY,TCHANO)
         CALL ZPUTTK(TNAME,LENGTH(NAME),NAME,TCHANO)
         CALL ZPUTTK(TZEOS,0,DUMMY,TCHANO)

      ELSE IF (EQUAL(SPEC,DP) .EQ. yes) THEN
         CALL ZPUTTK(TDOUBL,0,DUMMY,TCHANO)
         CALL ZPUTTK(TNAME,LENGTH(NAME),NAME,TCHANO)
         CALL ZPUTTK(TZEOS,0,DUMMY,TCHANO)
```

```
      ELSE IF (EQUAL(SPEC,LO) .EQ. yes) THEN
         CALL ZPUTTK(TLOGIC,0,DUMMY,TCHANO)
         CALL ZPUTTK(TNAME,LENGTH(NAME),NAME,TCHANO)
         CALL ZPUTTK(TZEOS,0,DUMMY,TCHANO)

      ELSE IF (EQUAL(SPEC,CO) .EQ. yes) THEN
         CALL ZPUTTK(TCOMPL,0,DUMMY,TCHANO)
         CALL ZPUTTK(TNAME,LENGTH(NAME),NAME,TCHANO)
         CALL ZPUTTK(TZEOS,0,DUMMY,TCHANO)

      ELSE IF (EQUAL(SPEC,CH) .EQ. yes) THEN
         CALL ZPUTTK(TCHARA,0,DUMMY,TCHANO)
         CALL ZPUTTK(TNAME,LENGTH(NAME),NAME,TCHANO)
         CALL ZPUTTK(TZEOS,0,DUMMY,TCHANO)

      ELSE

         CALL REMARK('Illegal Specification Removed:.',stderr)
         CALL ZPTMES(SPEC,stderr)

      END IF

      IF (COUNT .EQ. STOP) THEN
         RETURN
      ELSE
         COUNT = COUNT + 1
         GO TO 10
      END IF

      END

C     *** N A M E S ***

      SUBROUTINE NAMES(NUMB,PATH)
      INTEGER NUMB,PATH(*)

      INTEGER JUNK,PROMPT(21,4)

      INTEGER ZGTCMD
      EXTERNAL ZGTCMD,ZPRMPT

      DATA (PROMPT(I,1),I=1,20)/
     +      bigt,leto,letk,lete,letn,blank,lets,lett,letr,lete,
     +      leta,letm,blank,lparen,leti,letn,rparen,colon,blank,
     +      eos/,
     +      (PROMPT(I,2),I=1,20)/
     +      bigc,leto,letm,letm,lete,letn,lett,blank,letf,leti,
     +      letl,lete,blank,lparen,leti,letn,rparen,colon,blank,
     +      eos/,
```

```
     +          (PROMPT(I,3),I=1,21)/
     +           bigt,leto,letk,lete,letn,blank,lets,lett,letr,lete,
     +           leta,letm,blank,lparen,leto,letu,lett,rparen,colon,
     +           blank,eos/,
     +          (PROMPT(I,4),I=1,21)/
     +           bigc,leto,letm,letm,lete,letn,lett,blank,letf,leti,
     +           letl,lete,blank,lparen,leto,letu,lett,rparen,colon,
     +           blank,eos/

       CALL ZPRMPT(PROMPT(1,NUMB))
       JUNK=ZGTCMD(PATH,stdin)
       RETURN

       END


C     *** S T E P D I ***

       SUBROUTINE STEPDI(TCHANI,TCHANO,STATUS,TYPE,LEN,STR)

C Assume that the next token to be read from the token input
C channel TCHANI is either the first token of a program unit
C or the end-of-file token.  Read and examine tokens from this
C stream.  If the first token is an end-of-file, return with
C STATUS set to eof.  Otherwise, pass comments (except original
C directives) through to the token output channel TCHANO.  Examine
C the first pair of tokens of each statement to determine whether
C the statement is a program unit statement (a PROGRAM or
C SUBROUTINE or BLOCK DATA statement or a function subprogram
C statement), or an IMPLICIT statement.  If it is any of these,
C pass the statement to the output files.  If not, then, since
C we started the search at the beginning of the program, the
C statement is such that a type declaration may immediately
C precede it.  Do not output the statement but return the
C specification of its first two tokens in TYPE, LEN, and STR.

       INTEGER TCHANI,TCHANO,STATUS
       INTEGER TYPE(2),LEN(2),STR(maxtoklen,2)

       INTEGER BODY(81)

INCLUDE ZPTYPE

       INTEGER CHKINS
       EXTERNAL ZPUTTK,ZGETTK,CHKINS,ERROR

       SAVE

       STATUS = 0

C Read a token.  If it is the end-of-file, return with STATUS set
```

```
C to eof.  If it is a comment, pass to the output files (unless it
C is one of the original directives) and read another token; if it
C is not a comment, hold it and read a second token.

   10 CONTINUE
      CALL ZGETTK(TYPE(1),LEN(1),STR(1,1),TCHANI,STATUS)
      IF (STATUS .EQ. err .OR. STATUS .EQ. eof) CALL ERROR
     +    ('STEPDI: Error Reading Token 1.')

      IF (TYPE(1) .EQ. TZEOF) THEN
         STATUS = eof
         RETURN
      END IF

      IF (TYPE(1) .EQ. TCMMNT) THEN
         IF (CHKINS(STR(1,1),BODY) .EQ. no)
     +          CALL ZPUTTK(TYPE(1),LEN(1),STR(1,1),TCHANO)
         GO TO 10
      END IF

      CALL ZGETTK(TYPE(2),LEN(2),STR(1,2),TCHANI,STATUS)
      IF (STATUS .EQ. err .OR. STATUS .EQ. eof) CALL ERROR
     +    ('STEPDI: Error Reading Token 2.')

C The two tokens are the first two of a statement.  If the first is
C PROGRAM or SUBROUTINE or FUNCTION or IMPLICIT or BLOCK DATA, or if
C the second is FUNCTION, write the tokens for the statement (starting
C with the two in the buffers) to the output files and continue.
C (Note that the statement being transferred must have more than
C two tokens.)  If these conditions are not met, we have found a
C place to put specification statements - return.  TYPE, LEN, and
C STR have the right output values.

      IF (TYPE(1) .EQ. TPROGR .OR. TYPE(1) .EQ. TSUBRO
     +      .OR. TYPE(1) .EQ. TFUNCT .OR. TYPE(1) .EQ. TIMPLI
     +      .OR. TYPE(1) .EQ. TBLOCK .OR. TYPE(2) .EQ. TFUNCT) THEN

         CALL ZPUTTK(TYPE(1),LEN(1),STR(1,1),TCHANO)
         CALL ZPUTTK(TYPE(2),LEN(2),STR(1,2),TCHANO)

   20    CONTINUE
         CALL ZGETTK(TYPE(1),LEN(1),STR(1,1),TCHANI,STATUS)
         IF (STATUS .EQ. err .OR. STATUS .EQ. eof) CALL ERROR
     +    ('STEPDI: Error Reading Token When Transferring '
     +          //'Statement.')

         CALL ZPUTTK(TYPE(1),LEN(1),STR(1,1),TCHANO)
```

```
      IF (TYPE(1) .EQ. TZEOS) THEN
          GO TO 10
      ELSE
          GO TO 20
      END IF

ELSE

    RETURN

END IF

END
```

Open Forum

Chairman:	Wayne	R.	Cowell
The Panel:	Martin	A.	Broom
	Malcolm	J.	Cohen
	Martyn	D.	Dowell
	Stephen	J.	Hague
	Ian	C.	Hounam

and all other workshop participants

"I have a dream today.
I have a dream that one day ...
   ... little black boys and black girls will be able to
join hands with little white boys and white girls and
walk together as sisters and brothers."

— Martin Luther King,
Allocution at the March for Jobs and Freedom,
Washington, 28th August 1963

Prologue. From its inception, *"Using Toolpack Software Tools"* was thought of as a workshop rather than a course because of the intention of capturing and sharing previous experiences in installing and/or using the system, that the participants might have already gained. Therefore, we attempted to introduce tool usage with a practical flavour. Additionally, wide discussions were encouraged during the lecture periods and in informal groups at any available time. In conveying the workshop material to edited chapters, the authors have given consideration to relevant questions raised during their presentations. However, workshop sessions have become *chapters* and attendants' own points of view have been integrated into the broader context. Consequently, the spirit of open discussion cannot be perceived in the other chapters. An effort to maintain that spirit in full has been made in this chapter, intentionally entitled *"Open Forum"*, by reporting, in a lively and animated style, the panel discussion held during the workshop.

A. A. Pollicini (ed.), Using Toolpack Software Tools, 277–296.
© 1989 ECSC, EEC, EAEC, Brussels and Luxembourg.

Panel and General Discussion

Chairman : Over the past few days the panel has agreed upon six areas for discussion. This does not need to constrain you in any way. If you think of a question you would like to raise, you may certainly interrupt us and raise the question. For each of the six areas we will list several issues that will clarify what we hope to discuss in that area. We will then open the discussion to the members of the panel and hear from them; finally, for each of the issues we will invite everyone to participate in a general discussion.

The six areas pertaining to the proposed theme are these:

*Users' Requirements for a Fortran
Programming Support Environment*

- Acquisition and Use of a Programming Environment
- The Tool Set
- The User Interface
- Installation
- Documentation and Education
- Follow-Through Service

Chairman : The intention is to talk about programming environments in general, realizing that we are all interested in Toolpack in particular. Thus, the discussion may range further than Toolpack but each area has an interest within the Toolpack context.

We begin by considering:

Acquisition and
Use of a Programming Environment

- Application areas

- Size and type of programs

- Expectations of management

- Expectations of users

- Suppliers of environment software

  - Computer vendors

  - Software vendors

  - Public domain

Chairman : You have probably read in Toolpack project documents and in articles written about Toolpack, that at an early stage of the project, we formulated our objectives in terms of what kind of programs we intended to support. We aimed to support primarily people who were writing mathematical software in Fortran. We indicated the size of the programs that, we believed, were appropriate for that kind of support. But, in fact, Toolpack has been used, and we expect that it will continue to be used, in a larger context than mathematical software and applied to larger programs than the 10,000 lines of Fortran that we thought was a good size at the time. The size of programs and the application areas have been modified in the minds of the Toolpack developers as events unfolded and we expect better understanding of the applicability of Toolpack as we gain further experience.

To generate some discussion on the expectations of management, I have just passed to you the report *"Study of a Prototype Software Engineering Environment"* [49] from the National Bureau of Standards. *Management,* in this case, is the Government of the United States in its rôle of supporting large programming efforts. As outlined in this report, the National Bureau of Standards is developing techniques for evaluating programming environments and has applied these evaluation techniques to a pre-release version of Toolpack. (Although the report is useful for Toolpack developers and users, and generally gives good marks to Toolpack, it should be emphasized that the Toolpack evaluated was an early prototype.) There are many implications for this, but the question we are asking today is: what do the people who make decisions about what to be done in a large organization, expect of a Programming Environment?

What about the expectations of programmers? On page five of the mentioned report there is a table with a list of types of tools, a column that shows the importance, in terms of increased productivity, ascribed to that type of tool by programmers (the "Relative Choice"), and a column that shows Toolpack tools of the given type. (See below)

Tool	Relative Choice	Toolpack Tool
Interactive debugger	1.00	none
Screen editor	.98	*ISTED*
Subnetwork checker	.97	none
Process meter	.97	*NBS Analyzer*
Print file	.96	*ISTCI*
Stream editor	.91	ISTDS, *ISTGI*, ISTMP, ISTPT, ISTRF
Data dictionary	.90	none
Configuration manager	.90	ISTSV
Source code control	.90	ISTVC
Test coverage analyzer	.89	*NBS Analyzer*
Auto test generator	.88	none
Process monitor	.88	none
Private library	.88	ISTCE, *ISTCI*
Storage monitor	.88	ISTCE, *ISTCI*
File comparator	.84	ISTFD, ISTTD
Big file splitter	.84	ISTSP
Program cross referencer	.81	none
Display	.79	ISTCE, *ISTCI*
Big file scanner	.79	ISTGP
Source beautifier	.78	ISTPL

Tool Rankings from Hanson & Rosinski Study *

Chairman : The list of tools and the relative choices were part of a study by Hanson and Rosinski [27]. In that study a group of experienced programmers were asked to express their preferences for certain kinds of tools they need. The interactive

---

* Table reproduced from the report *"Study of a Prototype Software Engineering Environment"* [49], with kind permission of the Institute for Computer Sciences and Technology of the National Bureau of Standards.

debugger, the most commonly requested tool, is at the head of the list and the "Relative Choice" assigned to the other tools expresses the normalized preference.

We are posing questions like: What are the expectations of programmers with respect to tools? How many of these expectations will be met by a system such as Toolpack? How many will be met by computer manufacturers and software vendors? Who will supply integrated environments to meet the expectations of programmers? As a partial answer to the last question, we note that computer vendors supply compilers and interactive debuggers while public domain efforts such as the Toolpack project contribute tool suites. I tend to think the answer will be: "A combination of these". Let us ask if any member of the panel or audience would like to offer any idea about any of these questions.

Kvam : In real world usage, is the upper limit of 10,000 line programs a realistic one for Toolpack? Are "real Fortran programmers" not more likely to be working with 500,000 line programs? Can programs of this size be dealt with by Toolpack and/or is it realistic to try to handle such programs with Toolpack?

Chairman : You are pointing to the heart of the question. From the standpoint of mathematical software development, for example developing a package like EISPACK, it is reasonable to talk about 10,000 lines; but from the standpoint of people who have big simulation codes, for example, it is obviously too small. On the other hand, how often do you look at all the lines of a body of software? Do you look at a program unit at a time? At a few program units at a time? Perhaps 10,000 lines is quite large enough as a piece to be examined at any one time.

Kvam : It is true if someone is editing or working on the source code. However, if one wished, for example, to trace the use of variables within the flow of the program, then one would need to handle the parse tree of the whole program rather than just fragments. This will require that the tools are capable of parsing the whole program. Does the panel feel that there is a need for Toolpack to provide a development environment which enables it to handle large project development rather than individual codes?

Hague : Concerning what you said about "500,000 line programs", if you want to get an overall view, I suspect that nobody has an overall view of a half million lines, except in a very broad sense. They are probably the software suites that have the most need of tool support but, regretably they are least suited to Toolpack in the immediate future, since they are probably going to hit all the configuration limits of the present tool suite. Regarding project management, this is not an area that really has been centrally addressed in Toolpack. There are certain fragments of Toolpack which will be useful as part of larger project management schemes. That really has not been a major theme. So, I essentially agree with you, that it has not been fully addressed yet.

Chairman : Does either of you, or anyone else, believe that all the programmers' needs will be addressed by some one programming environment or it is more likely that some combination will be used. Toolpack may address some needs, compilers will address others, and interactive debuggers will address still others. Will we have to combine various kinds of resources to support the development of large programs?

Kvam :   I think of Toolpack as a facility that ought to exist in any type of Fortran programming environment. Concerning the expectation of the user and usually the expectation of managers, I think that, given a choice, they would prefer one package that would allow to perform all the phases of a job. Indeed, if a person tells you: "You don't need to do anything but work in a Toolpack environment because it is supposed to support every phase of programming", you say: "well, after a certain size the project is to big for Toolpack, so we are going to leave the Toolpack environment to go to some other tools". In these circumstances your feeling is: "Why was I wasting my time using Toolpack anyways if I need to go outside of it?" I mean to say, is there a need to try to expand the horizon of the Toolpack environment? To expand it not just in the sense that you are allowed to call up Fortran compilers from Toolpack or to have access to your favourite full-screen editor from within different tools.

Pollicini : I will address only part of that point. If a user is allowed to ask for much more than reasonably feasible, as a user I had a dream. This dream has roots in what is shown in Appendix E of the *"Tool Writers' Guide"* [32]; namely the example of how one can embed a Fortran compiler into the Toolpack tool suite. My programmer's dream is that, one day computer manufacturers will provide us with compiler fragments capable of producing and recognizing token streams, parse trees and symbol tables in the Toolpack format, so that Toolpack fragments and compiler fragments can share their interfaces and work together as co-operating units. It will then be possible to selectively enter the code generation phase (and perhaps the optimization phase) of the compiler by providing it with a parsed view generated by Toolpack fragments such as the lexer and the parser. Furthermore, it might be interesting to flatten and polish the optimized code, so that you have an idea of what language construct is actually executed. Will it be feasible, or remain a dream for ever?

Chairman : Does anybody else have any comments on his dream? If I understand what are you saying, the compiler would start from the parse tree and symbol table produced by Toolpack and would use other information produced by Toolpack.

Pollicini :   Yes, indeed, if we have tools which provide us with a parsed view of the source programs, in a form already suitable for code generation, and we feel this intermediate form is equivalent, if not superior, to the one produced by the compiler itself, we can consider that we will lose time if compilation starts by scanning and parsing once again our source program.

Chairman : Good. Are there any comments on any of these areas?

Detert : I would like to make a sceptical remark. Is it really feasible to provide an environment for Fortran programmers that is so complete that all typical applications or typical tasks a Fortran programmer normally does can be contained in such an environment? Fortran programmers, or maybe better software engineers, are not only dealing with a compiler, with maintenance of software; they are also concerned with text-processing, graphics and other tasks such as using networks and so on. At some stages, they will certainly have to leave the environment they are using, and this means that there is more than one environment. My impression is that it might be more useful to have a common set of facilities supplied by the computer manufacturers and then, for particular requirements, one can access special functions such as the ones provided in the Toolpack suite.

Chairman : To tailor the host environment. Yes, fine.

Hague : To make it functionally complete for certain tasks. This functionally complete environment is then made available across machines.

Chairman : Would you accept the compromise of having groups of people who are collaborating on a particular environment or are using the same environment?

Pollicini : We can certainly consider that compromise as an important step forwards in the spreading and consolidation of programming environments.

Chairman : Now we should move on because we have quite a lot of ground to cover, and I do not want to slight any of the other issues.

Tool Set Issues

- Priority of tool capabilities
- Extensibility of the tool set
  - by a vendor
  - by a publicly-supported group
  - by a number of contributors
- Modularity of the tool set
- Efficiency versus flexibility
  - "Flexibility" includes *portability* and *degree of modularity*

Chairman : Would anyone on the panel like to comment on any of these issues?

Hague : Let us take the first of those: *Priorities*. As usual with a collaborative project, the present set of tool capabilities reflect the prejudices and the pre-occupations of the most active participants. So the majority of the tools in Release 2 of Toolpack primarily reflect the concern of NAG, and they also reflect the interests of Argonne. But what should be done when we are entering into the new phase, where we can begin to think about reflecting the interest of a broader group? And maybe, one answer is to do what the NBS people have done, namely in some way to survey users' requirements. However, it is no use conducting a survey prematurely till people have some idea of what is possible or not possible. My own preference is to concentrate on the general transformational capabilities and also to put some efforts into translators, to translate various dialects into "Toolpack-Fortran".

On the extensibility question, I am quite sure that some vendors will, in some sense, make up their own set of tools and, providing they do their job properly, this may be highly desirable. I have the feeling that they might try to extend the subject language to their own dialect. I am not sure that this would be a straightforward step because they may not realize that the task is not as trivial as it seems. It is not just a matter of producing a new parser; semantically-based actions of the tools have to be changed to accept the additional constructs. So the implications of an extension of the dialect affects the whole of the tool set.

On the other hand, new tools contributed by Toolpack users within the present linguistic framework would be very welcome. I think the real problem there, is validation. Who will take the responsibility for validating what may be good, bad, or indifferent contributions? And these have to be collected together and re-distributed. What do the audience think? Do you want to see system-tailored versions of Toolpack (e.g. VAX-pack, Apollo-pack etc.) which accept the host Fortran dialect?

Kvam :   I have to agree. I think that one of the highest priorities is bringing more extended Fortran into the Toolpack domain. And I think that a further high priority would be to be able to take VAX Fortran with its extensions, maybe for special occasions a special parser and lexer for it. So you get it into the framework of Toolpack, where it should be possible to transform it into standard Fortran. VAX Fortran is very popular in the U.S.A., in both government and academic sites. These sites will always use VAX Fortran because they started and have continued to use all those extensions. I think that if Toolpack is going to gain wider acceptance we have to deal with those factors.

Another possibility is that vendors will choose to support Toolpack in its standard Fortran form, but they will adapt the tools to take advantage of the host system in some way. Users are likely to appreciate such vendor support.

Chairman : Questions of support are related to questions of who produces the tool set. This is the issue of "Follow-through Services" that we will consider a little later.

Verstappen : I think Toolpack is a very sophisticated programming environment, but I would like to have a more friendly user interface, a HELP facility.

Chairman : That is, indeed, a topic included in the next area and it is time to move to the next set of issues:

User Interface (Command Interpreter) Issues

- Level of sophistication

- Uniformness among installations

- Desirable features (e.g., HELP system)

Chairman : We thought of these as command interpreter issues in general, not in the sense of a particular command interpreter such as ISTCI, although that is a good example.

Broom : My opinion is that the HELP system at the moment is not good enough. I think it is extending, whether it needs extending by people contributing their ideas on how tools should be used. We have a HELP system too. You either extend the facilities of the use to users as tools are now, or you do it at each individual site which can be passed on by NAG to other sites.

Cohen : I have some thoughts on the level of sophistication and really I think one probably wants a variable level of sophistication instead of providing people only with incredibly sophisticated user interfaces. I can imagine rather a lot of Fortran programmers being turned off by that. So, thinking along the lines of ISTCI may or may not be a good direction in which to progress. But I think really a command interpreter needs to do things in a sophisticated manner. Toolpack does not really approach this problem; ISTCI is ultra-sophisticated and experimental, and most tools have only a very primitive interface themselves. I do not think that should be a problem in most installations because of the use of command files. Basically you want users to be able to say "Polish" and "DECS" and so forth, and it is not very difficult to make up simple command files to do this. The command files can then manage the intermediate files that get produced and so on, and the user really does not have to worry very much about that. I agree, there is a lot more work that can be done, but there is a lot more work that can be done on any user interface.

Chairman : Did you have some sense of how sophisticated you want the user/Toolpack interface to be? At one extreme is a command interpreter that manages all your files. You just say "I would like to have a view of a polished program", and the command interpreter determines what is needed and what exists. It then initiates processes that create the needed files. ISTCI is an experimental prototype of such an environment. At the opposite extreme is the initiation of the tools, one at a time, with the user managing all of the intermediate files. I suspect that many users would want a solution between these extremes. For example, as Malcolm Cohen points out, there is on Unix installations a set of command scripts, the kind of thing you can probably write on any system. Operations like polishing a program are accomplished by writing a simple command line. The script invokes the necessary sequence of tools, creating the intermediate files, but not making use of a data base of information about the program, as would a more sophisticated command interpreter.

Dowell : I believe that this is perhaps more of an area where one should accept that, in the future, individual implementations may wish to provide an environment in which the Toolpack suite is very much tailored to that computer system. If you start thinking that really you might like to use this sort of system on machines with interfaces which are mouse driven with pull down menus, then you can't avoid accepting that your user interface is going to be machine dependent. I believe that there are certain types of a user interface of this sort that are so interesting for a certain class of users that you have to allow that development to happen, and accept that if someone does a very good job on the interface to the environment of that type it will provide something which can be used much more efficiently or much more easily in certain circumstances. So I believe that this is more an area where you should accept divergence of product rather than in the area of tools for handling different dialects of Fortran.

Chairman : Yes, I suspect that we would inhibit development if we insisted that people running Toolpack on a sophisticated workstation had to limit themselves to the same kind of facilities as were available in the older systems on which it was running.

Defert : I think that this is a valid point because some organizations, for example, have committed to some user interfaces, and they are going to access all the software through the same user interfaces. For instance, in our institution, we are not going to re-write the command executor but to put it in another mode. In order to make it last a long time, there must be a commitment from Toolpack, not to change for example the communication between processes and between tools.

Chairman : That is an interesting point. I think perhaps we should ask Steve Hague if he wishes to say anything about that matter.

Hague : I think that at the heart of that question is the TIE definition or the inter-tool communication. There has always been a statement that at some point TIE would be revised and this was all part of the Release 1 material that said "Here is a discussion document about TIE 2". That is in the future. The second release of Toolpack/1 conforms to the original TIE. The desirability of re-defining TIE still exists and a lot of experience has been gained, and we stand by that assertion, that there is likely to be, there is intended to be, a revision of TIE at some later stage in the light of the overall experience gained. When that re-definition takes place, then that must be a firm, fixed line so that if institutions invest beyond that point they are secure.

Chairman : Do I understand you to be saying that we have not yet reached the point where we are ready to fix a definition but that we agree to the principle that such a definition should be fixed?

Hague : That is a correct interpretation of our current position. We are sure that the present tools are very useful but we caution sites against depending on those tools as the sole means of support for critical programming projects.

Chairman : Are there any other comments?

de Sadeleer : I would say that Toolpack could be very well menu-driven with a set of menus which respect well the structure of each tool, giving access each of them, to all the possibilities available. I think this is rather easily done, also in asynchronous mode.

Defert : There must be two levels of HELP. There must be a first level of HELP about each tool. And if you use the ISTHP tool, I do not think it gives you enough information in order to fulfil this former requirement. The second level is that contained in ISTCI but the command interpreter is not working necessarily on every machine. Perhaps, one might also think of some sort of expert system about Toolpack accessing information organized into files. I am sure you can enter very easily some information interesting for the users. There are two main areas of interest. There is the way you use the tools and there is the way you combine tools.

Chairman : Have you used the MAN facility in the Unix system? I wonder if anybody has any comments on that as a way to go.

Broom : Well, there is at the moment a facility quite similar to ISTHP, that is HELP which gives you a very small help, and there is MAN, which stands for MANual, where you get pages and pages. So I think, unfortunately, that the use of the MAN facility in Unix goes a little too far. There is a need to find a middle stage.

Chairman : I agree and I think the comment reflects that there needs to be a hierarchy of help information.

Defert : I think that, because there is only one level of HELP you can ask information about one tool or about one keyword and you just receive that order. Could it be like under the VMS system where the help information is structured? I do not think it would be very difficult to have more structured information available through ISTHP.

Kvam : Yes, I have to agree. I think, some kind of improvement should be provided, either something along the lines of VMS where you start with a general topic and then, if you are more interested, you can go into subtopics and so work your way down the tree structure, or even something in the way of NAG online HELP system, where you start out with certain things, and if there is more specific information needed, the HELP system could prompt you for going down deeper and give you better advise on different things. I think something like that could be helpful. I think also, for command oriented tools as the editor ISTED, if instead of just typing HELP and the only thing you get is a list of commands, if you could do something like HELP, let's say, APP and it would just give you a couple of lines describing what the command APPend does. In the present situation, with just a list of commands, if you have not got the documentation, it gives you no idea of what you are doing. Additionally, the tools would be a little more friendly if they could give you, for instance, the chance to say: "Oops I goofed, I want to go back". It would be very nice if that type of friendliness could be built in, which is something which could easily be done, I think.

Broom : I fully agree. Toolpack does not give you the chance of correcting your mistakes. In certain situations where you are trying to find a file, if the file is not there, the tool does not give you a second chance. With certain Toolpack tools, you enter a name, the tool goes and opens the file if it can, then it proceeds to the next point. In the case of other tools, they ask you for the names of all the needed files and then they go and look for the first one, so there is really an inconsistency there. The tool should ask for a name and if it does not find the file it should return and give the user a second chance.

Chairman : That would be an easy change to make.

Cohen : Actually I can explain exactly why that is not done and that is because the tools were meant to be able to be used in batch-mode. In such a working mode, if the tool comes back when it can't find one file it will retry with the next name and this will completely destroy your batch sequence, which is not a very good idea. And I take your point about wanting to go back and correct errors but I have yet to see very many processors or systems which let you do that. If you say: "Compile FRED" and you hit Return, the only way to stopping that on most systems, is in fact to stop the compiler and to say: "Sorry, I didn't mean FRED, I meant FRED2". You have to type it in again, going through the process of line editing and then recall the compiler having specified the correct name. I do not think Toolpack is particularly deficient there, compared with other systems I have used.

Dowell : I think that in the future we must give consideration to the problem of batch-mode and interactive mode. You have to be able to determine which mode you are using and to either take the batch-mode attitude towards command analysis, or the intereactive mode. Regarding provision of tools and HELP systems, one must consider the advantages/disadvantages of a Toolpack provided system against existing specific operating systems facilities and tools. Extending what Jeff Kvam has said about the VMS system, I can imagine that one may want to integrate the Toolpack tree of HELP into the existing branches of the VAX/VMS HELP and thus, what one needs is a hierarchically structured set of HELP data which could be integrated into an existing environment.

Hague : That is exactly the issue we are facing with the NAG HELP system. On VMS we are going to play it both ways. You can either use the NAG HELP system, a free-standing program, which you enter to obtain HELP about the NAG Library, or you have a slightly reduced version which you can embed in your native VMS HELP system. It is not quite as good; you lose some of the functionality, but it is more natural for the VMS user, and it is more efficient, and that is the trade-off.

Chairman : I think we can proceed and look at the next set of issues, which concern installation:

Installation Issues

- Flexibility available to installer

- Ease of installation
  - "Pre-fabricated" versus "on-site" construction

Chairman : It is not a long list of issues but it is a very hard subject. How much flexibility should be available for the installer? What about the ease of installation? I like to make an analogy with the building industry: On the one hand there are houses that are "pre-fabricated" in a factory and assembled on a foundation at the site. This process is more efficient but less flexible than "on-site" construction, in which the house is built from raw materials on the site. "On-site" installation of Toolpack starts with the basic elements (the "raw material") of Toolpack and creates a customized installation. If one installs a "pre-fabricated" version, such as is available for VAX/VMS or Unix, the installation is much easier but the user must be willing to accept the decisions of whoever prepared the installation.

Let's ask the panel if they have some thoughts on the subject of installation.

Broom : I think I can answer some of that concerns. My University was one of the first sites to get Toolpack in its original form. I must admit I spent probably a week to sort out the documentation and tried to install Toolpack without much success. I think, if you take a version of Toolpack already implemented on a specific system, it is very straightforward. What you have to do is to physically produce your main files for the new system, starting from the NAG tape which contains the implementations already available on some machines. So, if one such implementation meets your requirement, this is the advisable solution. It is a lot of hard work if you do not know that. I think there are a few people here that installed Toolpack on other machines, I am sure they encountered many problems doing that.

Defert : I installed it on a VMS machine so it was easy but it took me about three weeks to make it working because there were some errors and I had to debug each time the software failed. However, I must agree that I learnt a lot of things just by customizing the software to my operating system, just to know more about the structure of the TIE library. For Release 2 I have already ordered a tape. I have preferred the option offered by the "Service C", that is the VAX/VMS specific version because I do not want to go a second time through the whole customization process. However, I do not regret to have done it once, because I have learnt something.

Chairman : I understand that NAG and the NAG User's Association (NAGUA) are thinking of organizing a one-day workshop that will address the question of new installations. Is that right? Could you respond to that, Steve?

Hague : We haven't fixed the meeting in every detail yet, but I can say broadly what it will be like. It is going to be a one-day event, probably in London. *

If it is successful we may consider holding similar events elsewhere in Europe, and the idea would be that it would be primarily for people to come and have some general introduction to the installation process and then have what we might call a "walk-through" installation, during which you observe a lecturer go through the tool installation procedure. This is all artificial: it is a little like a T.V. cookery-program where in twenty minutes they produce a beautiful dish that takes hours to produce. But at least you'd have a rehearsed example and then perhaps we envisage that the meeting might split up in parallel sessions if there is enough interest. There may be a group of people interested in IBM specific problems, for instance, and we would try to encourage people who have got at least some experience on those machines to come and act as the focal point of each of those parallel sessions.

Kvam : Regarding the installation of Toolpack, holding the course is a very good idea that would help a lot of people. At least in the United States there is more

---

* The Toolpack/1 Implementations' Workshop was in fact held on 3rd July 1987 in London.

of a need. People really just will not do the work or would give up and there have been very qualified people who have not been able to do that. Now I was wondering if it would be possible, let's say, not going into the realm of making Toolpack really a product and keeping it public domain but at least to charge a slightly higher fee for the work of making it run on a different type of machine. Say, like someone who wants the tool suite implemented on a Harris system, consulting fees can be charged if that people would not have to go through the troubles of an implementation effort.

Detert : I would just like to give an idea of what it means to implement TIECODE and some tools under the IBM systems VM/CMS. I have some figures about what was required to really install it.

I had to change 52 split files for customization to VM/CMS and to apply update notices. Most of this work concerned the application of changes reported by the update notices and thus it was common to every installer of TIECODE Release 1, irrespectively of the target system. The harder problems, hovever, were caused by changes concerning the customization for VM/CMS. Additionally, 18 command procedures (REXX EXECs in VM/CMS glossary) had to be provided. Five of them for implementing functions lacking in IBM Fortran, such as dynamic allocation of files and inquiries about file existence and status. In particular, in the VS FORTRAN compiler, the INQUIRE statement is not implemented as specified in standard Fortran and, therefore, it does not behave as Toolpack assumes it should do. These command procedures are called from within TIECODE and provide access to operating system functions. Eight other were needed for calling Toolpack tools and for accessing run-time libraries, e.g. to emulate a hierarchical library structure. Five more were provided as installation aids for performing such operations as compile, split, compare files, search duplicates in libraries, format and print documentation. Furthermore, 10 Fortran routines had to be completely re-written (bitwise logical operations, time and date, CMS naming conventions for files etc.) and 9 editor macros had to be written as installation and documentation aids.

This is really a great amount of work which had to be done with Release 1 and we feel that a similar situation is not acceptable for the future.

Chairman : On that list of things that you did there are some that the installer is expected to do, such as writing time and date routines. The specifications of these routines are given. There are other things on your list that depend tightly on your host system. Do you have a rough estimate of man/months required to install Toolpack on your system?

Detert : My estimate is about three man/months.

Chairman : As the administrators in your organization know, one fourth of a professional man/year is a substantial amount of money.

Detert : We would be willing to pay a substantially larger amount of money for a version which is really working rather than to begin such an installation and not knowing when we will be finished.

Pollicini : I have just one comment on the inter-dependence between "Flexibility available to installer" and "Ease of installation". I would like to warn prospective installers about the fact that, by using a pre-fabricated version one looses the flexibility offered by customization. Conversely, if one wishes to customize his implementation for its own site, there is little opportunity of a quick installation which is only allowed by copying ready to use files. On-site customization has some intrinsic advantages, I just mention one parameter: *buffer-size* which influences the physical number of I/O and consequently the actual performance of the tools.

Chairman : We move on now to the area of:

Documentation and Education Issues

- Documentation
  - For users
  - For installers
  - For tool-writers

- Courses/Seminars/Workshops
  - On using Toolpack
  - On developing new installations
  - On writing tools

- Tutorial material (e.g., books)
  - On using Toolpack
  - On writing tools

Chairman : These issues relate to the dissemination of information about the installation and use of Toolpack, and about the writing of new tools for Toolpack. This workshop on using Toolpack has been very successful. Other areas that could be considered in similar courses are "New Installations" and "Writing Tools". The proposed NAG/NAGUA seminar on installing Toolpack was noted above. What sort of tutorial material do we need? How much interest is there in writing new tools within a Toolpack framework? I want to ask the panel first, if they have any comments about these issues.

Dowell : This is another area where you cannot solve the whole problem all at one time. The first problem that we have seen manifesting itself in the last one or two years, is getting Toolpack installations into a working state. So, in the area of installation, steps have been taken to provide a better understanding of the problems, even perhaps a better understanding of what needs to be done to make the problem simpler. Concerning Release 2, certainly documentation for installation looks much better and much more useful for installation purposes. The idea of some sort of workshop or course for people who are installing the system is a good one.

In parallel, you also have to address the problems of the other people who want to use that software once it is installed. That is an area, where we, at Ispra, are now looking at what we are going to provide for our users. We have to start to promote Toolpack for our users and provide them with sufficient education and documentation to put them into a position from which they can take profit of the tools in practice. The first step is that Toolpack has to get off the ground, it has to convince more people to install it. It has to be simpler to install. The actual installation process has to be made clearer to people by various means and then, in term of usage, you have to provide the sort of courses and the sort of documentation that end-users can benefit from. That means that courses have to be fairly straightforward and somewhat less technical.

Chairman : Anyone else on the panel? Aurelio, do you have any comments as the organizer of this workshop?

Pollicini : The organization of workshops, practical sessions and the like is certainly an overwhelming task, however, similar initiatives should be repeated because user education is essential in the promotional phase, especially to help the user in getting the most out of the system. What is also essential is to provide the appropriate support documentation because *products are successful if documentation is good.*

For instance, in the context of what Martyn Dowell has just said, it is our inention to organize a seminar on the use of Toolpack tools as part of the computing service of JRC Ispra. We will carefully think of the scope of the local Toolpack service and of the requirements of our end-users before preparing the documentation specific for our host system. Indeed, we feel this is important for achieving both widespread use and users satisfaction.

Concerning books, it is planned to have the proceedings published for this workshop. In this case, of course, the covered field is "Using Toolpack Tools". "Writing Toolpack Tools" is a second topic for which a book would be very welcome.

Chairman : We come now to questions about:

Follow-Through Service Issues

- Maintenance of software

- Coordination of development

- Sponsorship of educational activities

- Information exchange among users

Chairman : Since NAG is very active in these areas, let's see if Ian Hounam and Steve Hague could start us off on this one.

Hounam : I would like to tell you about Toolpack maintenance of software. We prefer to receive written bug reports but often we receive them by telephone which is very time-consuming. In the future we will still be prepared to receive bug reports, and if you subscribe to one of our services, namely the "Information Service", you will receive both bug reports and newsletters. We cannot guarantee to give you a quick response, but you will get a response if you subscribe to this service. The newsletter also provides another follow-through service, that will be news of any service updates, new tools that have been contributed and new TIECODE installations that were contributed.

Hague :   It would be inappropriate to go too far into internal NAG policy matters. However, the question of support services for Toolpack is a very relevant issue because people, particularly in North America, have gained an adverse impression of Release 1, and we are anxious to counter that in Release 2. It is worth saying a little about NAG involvement here. NAG, traditionally, has been a software library producing organization. And for some of the time of our involvement in Toolpack from the late 1970's onwards, we discussed from time to time what proportion of our limited resources should go into library development and what should go into software tools. Now that is no longer debated. All our colleagues working on libraries are convinced of the necessity of these tools for their own work. We also recognize that the tools have commercial possibilities in their own right and that, in the longer term, the experience gained in distributing interactive tool suites will help us build environments in which our other products can be better presented. Thus, there are three reasons for our involvement : our own use, tool services, and gaining experience of building portable products for scientific environments. Now it is all a question of pace. We need to listen to what people say and then judge for ourselves where to put our effort.

My own view is that on those machine ranges where we have the required expertise, resources, and experience, we have already started to provide enhanced services. It is evident in the Release 2 description that we are beginning to step up the level of service. We are sufficiently confident about VAX/VMS that we are going to stand by the assertion that you can read a tape to install the system in the time it takes to read the tape. The Unix assertion will be almost as strong but not quite. I think that's the evidence of our seriousness. We are beginning to increase the level of service where we know we have the resources and experience to do it. The question of services on the other machine ranges is really matter of trying some experiments with individual customers to see what is involved in assisting them to install these tools. We would hope to be able to give a more complete answer in about twelve months' time, when it should be clear what level of general commercial tool service, in addition to enhanced VMS and Unix services, will be available from NAG. At this time it would be inappropriate to launch a general multi-machine range supported tool service. What is in the public domain remains in the public domain. What we are talking about here are services built on top of the basic system.

Chairman : I will re-affirm what both Jeff Kvam and Steve Hague have said about public domain software versus commercial software in North America. Generally, outside of research settings, there is less interest in adapting public domain software. The preference is to buy software and hold the vendor responsible for it. I suppose that there are a number of social reasons for this. A major factor, I believe, is that, in Europe, there are more good people working at adapting public domain software. In the U.S., many of the equivalent people are working for companies that produce commercial software.

Kvam : I would like to give you an idea of the difficulty with Release 1 that we have in North America. In the U.S. office of NAG Incorporated we have distributed approximately 200 copies of Toolpack. To our knowledge there are less than ten sites that have Toolpack in use at the moment, because of this installation difficulty and the associated expense in manpower. With the Unix version and the VAX/VMS version, which are easier to install, we would be able to address in the order of about half of the cases. That's why I think it is important to distribute such versions.

Hague : I like to make a couple of points. One is, in many ways I am full of admiration for what my colleagues at NAG, Bob Iles and Malcolm Cohen, have done: they carried out a tremendous design and programming job in a very short time. On the other hand, we know that difficulties encountered by some Toolpack installers have adversely affected NAG's reputation. In part, that is because when they saw the word "NAG" they assumed that the tape they receive would work after, at most, a couple of days' work. My second point is about the possible state of Toolpack in two years' time. Let's suppose that there is a public distribution service and a commercial service from NAG and maybe others. Could these two forms of service co-exist? The services would have a common core of material of, we trust, a uniformly high quality. The commercial services would then have perhaps greater functionality or a more specific machine-orientation. Also, we can envisage a situation in which people will still contribute material which would go into the public domain release but in some circumstances we might then ask the contributor if we could also introduce it into the commercial release. In this way we envisage that two different valid and useful forms of service can co-exist.

Chairman : It is very pleasing to have had such a wide-ranging discussion with such a constructive remarks from so many participants. However, the allotted time has elapsed and so, with my thanks to panel and audience, I now close this session.

Epilogue. In editing the discussion recorded during the panel session, I had recurrent occasions of *walking-through* the many subjects which had been raised. Therefore, I feel committed to append a few concluding remarks. In particular, I would like to focus on three emblematic points emerging from the discussion. Firstly, the idea that marketed, i.e. professionally developed and maintained software must be preferred to pieces of *do-it-yourself* software, seems finally to be prevailing also in the field of scientific applications. Secondly, the integration of a transportable tool-kit, such as Toolpack, into a host environment is advocated both as a whole and at the single facility level – e.g. the HELP facility. Finally, the definition of user interfaces is the area – I am tempted to add the *only* area – in which a certain degree of machine dependence is in the users' interest.

The Future of Toolpack/1

Ian C. Hounam,
NAG Ltd
Oxford, U.K.

*"In research the horizon recedes as we advance,
and is no nearer at sixty than it was at twenty."*

— Mark Pattison,
Isaac Casaubon, ch.10, 1875

Abstract. The enhancement of Toolpack/1 may be made in a number of ways:

Functionality,

NAG internal use,

Language domain,

User interface,

Portability,

Support service.

Improvements may be made in any or all of these directions and this chapter is devoted to a discussion of the possible and probable directions as perceived by NAG Ltd.

1.   Functionality

An obvious way to extend the functionality of the tool suite is to increase the number of tools available. This may come about in a number of ways.

(1) The contribution of new tools. As has been explained in the Chapters "Structure of Toolpack/1 Software" and "Tool Writing", the development of new tools using the Toolpack/1 conventions can be achieved with much less effort than would be required for a self contained tool. In the past this has resulted in tools such as ISTDC (see Chapter "Documentation and Non-Fortran Tools") and ISTUD (see Chapter "DO Loop Transforming Tools") being contributed to the tool suite. It

A. A. Pollicini (ed.), Using Toolpack Software Tools, 297–302.

is hoped that any new tools written by Toolpack/1 users will be passed on in this way in the spirit of public domain software.

(2) Tool developments at NAG, see Section 2.

(3) Developments in the target language will change the tool functions necessary to support the language. This is discussed in Section 3.

1.1.   The NEWTON Debugging System

A perceived need for a dynamic debugging system that is portable across a range of machines resulted in the NEWTON system that was distributed at Release 1. Because of a number of limitations in this software this has been withdrawn from Release 2 and the programs have been rewritten, but are not completely finished. The two tools associated with this system have been included on the Release 2 tapes, but for information only. Missing, at the time of writing, is the run time library that provides for user interaction with the instrumented program.

The NEWTON system will provide all the facilities of a dynamic debugger, but in a portable Fortran 77 format. These features are:

(1) Breakpoints.

(2) Single statement execution.

(3) Print out of the value of any variable.

(4) The user can change the value of any variable.

(5) Execution trace, by statement or program unit.

The NEWTON Instrumentor

The above facilities are provided by instrumenting the user's program with calls to a run time library that handles breakpoints, tracing etc. At the entry to each routine a call is made to a routine that will handle the tracing of the flow of control between program units. Before every executable statement a conditional call is made to the run time routine that handles the printing out and setting of the values of variables. This is only done if a breakpoint has been set or the single step execution flag is true. A call is also made to the run time library at the exit from every program unit. This allows the reporting of the statement number and type that caused the exit to be made.

The value of all variables is available, for examination or change at breakpoints, through the creation of new common blocks by the instrumentor. Hence the trace of program unit invocation may include, for example the values of subprogram parameters. The user will be able to view the current value of any variable and, if necessary, change any value. In order to allow access to DO loop control variables the NEWTON instrumentor rebuilds all DO loops with assignment statements and an IF – GOTO statement equivalent to the terminating condition.

The specification for NEWTON allows for separate instrumentation of program units. Hence the loader is required to build a global symbol table for the run time look up functions, combining all the separate program units.

The instrumentor tools have been written, but at the time of writing the run time library is incomplete. The Release 2 tapes will contain these tools, for information only. No time scale can be given for the completion of this project, but interested users will be kept informed on progress.

2.   NAG Internal Use

Within the NAG organisation extensive use of tools from the Toolpack/1 suite is made in the development and maintenance of numerical software. This has three main benefits to Toolpack/1 users.

(1) The tools receive extensive testing before release.

(2) It is not always easy to imagine all the potential uses for a tool and often discussion with active users of a developing tool can lead to simple extensions that increase the functionality of the tool.

(3) Suggestions from NAG internal users of the tool suite have led in the past to new tools being written in order to meet internal needs, for example ISTME (see Chapter "The Toolpack/1 Tool Suite").

Such feedback from NAG colleagues is expected to lead to new tools. For example, at the time of writing a prototype tool for the automatic generation of outline documentation of subroutine/function calls for a subroutine library is being developed. If these new tools prove to be of interest to Toolpack/1 users then they will be added to the tool suite at a future release.

3.   Language Domain

Evolution of Toolpack/1 in terms of the Fortran language that is acceptable to the tools has taken place from strict ANSI standard Fortran 77 to the Toolpack/1 Target Fortran (described in Section 4 of the Chapter "The Toolpack/1 Tool Suite") that allows a selection of commonly found extensions to the standard. These extensions mean that the needs of most users are taken care of, but some potential users of Toolpack/1, who program using the non standard extensions to Fortran allowed by some compilers, are denied use of the tool suite. Further language developments could be envisaged to support specific dialects of Fortran 77 but there are no plans for this at present.

The Fortran language itself is evolving and in the near future a new ANSI standard will define a new generation of the language called Fortran 8x [2]. It is unlikely that the language will be in common use until well into the next decade, because of the dead time in producing new compilers. The publication of a definite standard should signal the start of effort into producing tools to support the new generation of the language. New features likely to be in the new standard will have a profound effect on the tool suite. The intriguing possibility of tools to convert between Fortran 8x and Fortran 77 where this is possible should be explored. This could allow development of Fortran 8x programs to take place in advance of compilers for the new language.

NAG Ltd and the University of Salford, at the time of writing, have just started a collaborative project sponsored by the UK government Alvey Directorate, part of which is to work on tools for Fortran 8x.

4.   User Interface

Another aspect of the Toolpack project that could be developed is that of the user/tool interface. Current interfaces are ISTCE and ISTCI (see Chapter "Toolpack Invocation Techniques").

The current status of ISTCE is that it provides a working interface to both the tools and the Portable File Store and little development can be imagined in these areas. However, this interface might be extended in some implementations to allow access to the Fortran compiler from within the closed Toolpack/1 environment. This may not be possible on all systems.

ISTCI is still an experimental tool. Although further work on the Odin command interpreter, on which ISTCI is based, is taking place at the University of Colorado at Boulder, no further work on ISTCI is planned at the present.

5.   Portability

Toolpack/1 is portable across a wide range of machines, but the current portability base has proved to be more difficult to implement on some types of machine than on others. Hence developments here will concentrate on making the portability base easier to implement and also more efficient. In order to achieve this goal a new version of the portability base is being developed.

5.1.   A New Definition of TIE – TIE/2

The experience of putting together the first release of Toolpack/1 and writing new tools for Release 2 has influenced thinking on the definition of TIE. A major rewrite of the tool suite, for example, to move to the next Fortran standard, will be accompanied by a redefinition of the portability base. Some of the major points that will be addressed are:

(1) To provide a smaller definition more suitable to the tools and command interpreter that have been added for Release 2.

(2) To ensure that the definition is easy to implement, and that the resulting implementations can be reasonably efficient.

(3) TIE implementations should be possible over a much wider range of computers, including larger personal computers.

(4) To allow continued expansion through the use of supplementary libraries and easier integration with existing library facilities.

The intention is that the portability base should be versatile, and efficient and must be easy to use, install and implement. These requirements point to a much smaller and tighter definition than was provided in TIE/1.

Great care will be taken to separate the machine dependent code into a set of primitives on which the rest of the code can be built. The installer will be presented with these "Level 0" routines as those he must modify and often rewrite to install TIE/2. The rest of the code "Level 1" will be made up of completely portable code and calls to these primitives. "Level 1" code need not be modified for installation, but might be to aid efficiency.

Of the four possible TIE regimes, it has been found that the most useful are the Stand-Alone and fully Embedded regimes. Only these will be implemented in TIE/2.

The new definition will make it possible for the tools to be operated on by TIE/2 conforming tools. The consequence of this is that there will be no macro preprocessing or include files. The macro name facility of TIE/1 will be replaced by PARAMETER statements, which can be maintained by a tool like ISTPP.

This short introduction to the definition of a new TIE is intended to give an impression of current thinking. It is not a concrete definition. Work is currently underway at NAG Ltd to develop and implement this new definition as part of an Alvey sponsored project (see Section 3).

6.   Support Service

Currently Toolpack/1 is public domain software and is distributed by NAG in that spirit.   The concept of this kind of software in certain quarters is found to be unacceptable. In the commercial world what is often required is the "turnkey" system where the software vendor is often responsible for installation and maintenance. The idea that the software purchaser should have to invest effort in tailoring parts of the system is unheard of.

On the other hand in the academic world the idea of inexpensive software in source form that can be tailored to individual needs is more appealing. The user is quite pleased to make the effort of learning about a portability base and augmenting the software by writing his own programs. All this effort can be exchanged and information interchange by bulletin board or information sheet can put him in touch with others with the same interest.

These are two contrasting and very general views of public domain software. At the present within the public domain concept Toolpack/1 can only partly satisfy the needs of the first hypothetical user. Easy installation can be achieved for a limited range of machines:

(1) The machine specific implementations, and

(2) Some of the contibuted TIECODE installations.

It is in the interest of furthering the aims of the Toolpack/1 project to increase the number of such implementations. Hopefully installations of TIECODE on more machine ranges will be contributed. NAG Ltd would also like to get involved in projects that might lead to machine specific tapes for further systems. There are no definite plans, at the time of writing for any specific system.

The support of NAG Ltd for the public domain distribution service for Toolpack/1 will continue for at least another year, but during 1988 this may be reviewed. What is certain is NAG's involvement in software tools in support of the Fortran programmer but not what form this will take. It is likely that this support will gradually take a more commercial form. The change will not take place at once, nor will Toolpack/1 be withdrawn from public access. What is likely is a gradual separation of the public domain distribution service from commercial services at first based on Toolpack/1. In the end it would be expected that there will be fully supported NAG Fortran tool products.

Appendix A
Toolpack/1 Contents Summary *

N A G - Numerical Algorithms Group

CONTENTS SUMMARY
TOOLPACK/1 (RELEASE 2)
NP1366

The contents of the tapes distributed under the NAG Toolpack/1 (Release 2) services A, B and C (as described in the Service Summary) are as follows:

(A) Basic Distribution Service

Sites ordering Service A will receive a magnetic tape in the requested format and one copy of each of the printed documents listed below. The contents of the tape are detailed below. All tapes are written as 9-track, phase-encoded, 1600 b.p.i. containing upper and lower case characters and including the following ([] ! | \ ^ ' ~). The tape formats available are as follows:

Service A.1 - unlabelled ASCII, 80 bytes per record and 4000 bytes per block.

Service A.2 - unlabelled EBCDIC, 80 bytes per record and 4000 bytes per block.

Service A.3 - labelled to ANSI X3.27 - 1978 level 4 at 80 bytes per record and 4000 bytes per block. This format is largely compatible with VAX/VMS COPY format.

Service A.4 - Unix (4.2Bsd) tar format.

The details below indicate not only the functionality of each item but also indicate where changes have been made since Release 1.

Printed Documentation

One printed copy of each of the following documents is provided. With the exception of the Introductory Guide, these documents are also provided in machine readable form on the tape.

- Toolpack/1 Introductory Guide (Edition 2)
- README Guide
- TIECODE Installers' Guide

---

* This Nag Publication is reprinted with kind permission of Nag Ltd.

A. A. Pollicini (ed.), Using Toolpack Software Tools, 303–309.

- TIEVMS Installers' Guide
- TIEC Installers' Guide
- Tool Installers' Guide

TIE Implementations

The following implementations of TIE are provided on the tape:

TIEVMS - An implementation of TIE in a mixture of Pascal and Fortran 77 for DEC VAX/VMS sites. Since Release 1 this implementation has been improved. The .DIR empty file is no longer required and arguments are picked up automatically from the command line, if provided. A Pascal compiler and a 4.2 version (or later) of VMS are required.

TIEC - An implementation of TIE written in C for Unix based systems.

TIECODE - A transportable implementation of TIE written in Fortran 77 is provided which may be installed on computers other than VAX/VMS and Unix systems. For Release 2 TIECODE has been modified so that BLOCK DATA is no longer required.

Supplementary Libraries

ACCESS - The ACCESS supplementary library contains functions for accessing, modifying and writing the intermediate objects (such as parse trees and token streams) created by Toolpack/1 (Release 2) Fortran processing tools. This library has been extensively modified. The access functions for ISTPR format trees have been removed and additional functions to allow access to the attribute information generated by ISTSA have been added. There have been extensive changes made to both the formats of intermediate files and the operation of token stream access functions to improve the performance of the tools. There is no longer any BLOCK DATA in this supplementary library.

STRING - The STRING supplementary library contains additional routines for handling strings, in particular the pattern matching and replacement functions. A new routine has been added to check the legality of Fortran variable, routine and common block names. There is now no BLOCK DATA associated with this library.

TABLES - The TABLES supplementary library provides facilities for managing lists, queues and table structures. A new set of linked list functions has been provided.

WINDOW - The experimental WINDOW supplementary library provides screen and window handling facilities. Pop-up menus and information windows have been added.

Tools

ISTAL - A documentation generation aid for creating formatted reports from the information derived by other Fortran analysis tools (ISTAN and ISTYP). Improvements have been made to the type of report generated.

ISTAN - A Fortran 77 instrumentor for the dynamic analysis of software. This tool allows execution and control flow statistics to be gathered.

ISTCB - A new tool to perform differencing between two text files. This tool marks the changes with ISTRF formatting commands so that change and delete bars can be added to documentation.

ISTCD/ISTSB/ISTUD - A set of three new tools for unrolling and condensing DO loops. The restructured code is in a form that can be more easily optimised by compilers for vector machines.

ISTCE - A simple command executor tool. This tool recognizes the Release 2 tool names.

ISTCN - A simple, stream based token stream editor that can be used for changing names, strings or comments; a new tool at Release 2.

ISTCR - A name changer for Fortran program units that works selectively on the symbol table; a new tool at Release 2.

ISTDC - A data file comparison tool for comparing files of results.

ISTDS - A declaration standardiser that rebuilds the declarative parts of Fortran 77 program units according to a programmable template. This has been extended, both to handle new data types (e.g. INTEGER*2) and also to allow the use of INCLUDE files (see ISTIN).

ISTDX - A small tool to extract marked comments from program units and put them in a form suitable for the text formatter (ISTRF).

ISTED - A portable line-based editor. The text buffer may be analysed by the Fortran 77 scanner.

ISTET - A new small tool for expanding tabs in a file.

ISTFD - A simple Fortran 77 differencer for detecting changes between Fortran 77 files at the token level.

ISTFI - A small utility for finding include-file dependencies.

ISTFL - A small file length utility and example tool for the Tool Writers' Guide.

ISTFP - A fast unscanner that deletes comments and uses fixed spacing on tokens.

ISTFR - A new tool to convert REAL, DOUBLE PRECISION, COMPLEX and DOUBLE COMPLEX constants to a consistent form.

ISTGI - A tool to turn intrinsics into generic references. Since Release 1 this tool has been modified to accept the defined DOUBLE COMPLEX intrinsics.

ISTGP - A generalised pattern matching utility.

ISTHP - A simple help program.

ISTIN - A new tool to permit the use of INCLUDE files with the declaration standardiser (ISTDS).

ISTJS - A new tool to manipulate FORMAT statements, can join adjacent strings and turn H and X edit descriptors into strings.

ISTLS - A long name changer. This new tool allows Fortran programs to be written using long variable names that are then mapped to legal names.

ISTLX - The Fortran 77 scanner tool. This tool has been extensively modified since Release 1. The tool can now split up files into individual token streams, one per program unit, to assist in the processing of large input files. The error file has now been dropped and the list file made optional. There are changes to the format of the token stream and comment file to speed operation. The grammar has been modified to handle all types of comments, to allow DOUBLE COMPLEX and to improve general processing.

ISTME - A new tool to modify complicated expressions so that during calculation the sub-expression stack depth is kept to a minimum.

ISTMF - A new Fortran-sensitive differencing program that will insert the differences as comments in the output code.

ISTMP - A general purpose macro processor.

ISTPF - A major new tool that provides a Fortran 77 equivalent of the Bell PFORT portability verifier (which was for Fortran 66).

ISTPL - A source code polisher or pretty-printer. Several new options have been added.

ISTPO - A menu-driven option editor for ISTPL.

ISTPP - This tool ensures the consistency of PARAMETER statements between program units.

ISTPT - An arithmetic precision transformer which can convert Fortran program units between single and double precision. This tool has been modified to handle the defined COMPLEX/DOUBLE COMPLEX mappings.

ISTRF - A text formatter similar to the Unix 'roff' facility. This tool has been modified and extended. In particular, preprocessing of documentation through ISTMP is now no longer required.

ISTSA - A major new tool that provides a semantic analysis capability to check conformance to the Fortran 77 standard .

ISTSP - A file splitting utility.

ISTST - A Fortran 77 structurer. A major new tool similar in design to the Unix 'struct' facility but producing Fortran 77 output (from Fortran 77 input).

ISTSV - A save/restore tool for the portable file-store.

ISTTD - A simple text differencer tool.

ISTUN - A new small tool to undo the effects of ISTIN after the use of transformation tools.

ISTVA/ISTVS/ISTVT/ISTVW - A series of tools that allow the user to view the intermediate objects created by tools such as ISTSA and ISTYP and may also be used to detect some minor coding problems.

ISTVC - The version control tool.

ISTYF - The parse tree flattener.

ISTYP - The Fortran 77 parser. This tool has been modified to speed up operation.

Combined Tool Fragments

The following tools are all new at Release 2 and provide combined (and much faster) forms of a number of commonly required tool sequences.

ISTDT - The combined ISTYP/ISTDS/ISTPL tool; token stream to source code with declaration standardisation and polishing.

ISTLA - The combined ISTLX/ISTYP/ISTSA tool; source code to full static analysis.

ISTLP - The combined ISTLX/ISTPL tool; source code to source code polishing operation.

ISTLY - The combined ISTLX/ISTYP tool; source code to parse tree.

ISTP2 - The combined ISTLX/ISTPP/ISTPL tool; source to source PARAMETER value setting.

ISTQD - The combined ISTLX/ISTYP/ISTDS/ISTPL tool; source code to source code declaration standardiser and polishing.

ISTQP - The combined ISTLX/ISTYP/ISTPT/ISTPL tool; source code to source code precision conversion and polishing.

ISTQT - The combined ISTLX/ISTYP/ISTPT tool; source code to token stream with a precision conversion.

Experimental Tools

The following tools are currently provided in experimental forms only and should not be considered to be fully functional.

ISTCI - An experimental object orientated command interpreter.

ISTJF - A Fortran 66 source level differentiator that will create a derivative function from the analysed source.

ISTX1/ISTX2/ISTX3 - Three small experimental expert systems written in Fortran.

NEWTON - A run-time interactive debug system. This software is currently incomplete but will be further developed during the life of Release 2. The Release 1 Newton system (ISTPR, ISTNA, ISTNI, ISTNL and ISTNR) has been deleted.

Documentation

Documentation is provided in machine readable form for all tools. In addition there are a number of general documents that describe aspects of using Toolpack/1 or writing tools for use within Toolpack/1. Printed copies of these documents may be obtained under Service E. The machine readable documents need to be processed by the Toolpack/1 (Release 2) text formatter (ISTRF) before printing. The documents may be processed using the Release 1 tools (ISTMP and ISTRF), though the results will not be as inteded.

(B) TIECODE Distribution Service

Sites ordering Service B will receive one or more copies of TIECODE adapted for specific host environments. The tapes available at any given time will increase in number as new contributions become available. In general, each tape will include customised forms of the SPLIT and TIEMAC utilities, an installed version of TIECODE and some documentation. The tape may be in a host specific format. Additionally, tapes may contain script files to assist in the installation process or the use of tools. Sites ordering this service should note the limitations of support detailed in the Toolpack/1 (Release 2) Service Summary.

Service B.1 - IBM MVS/TSO

Service B.2 - IBM VM/CMS

Service B.3 - CDC Cyber NOS-2

Service B.4 - CDC Cyber NOS-VE

Service B.5 - Harris Vulcan

Service B.6 - DEC 10 TOPS10

(C) System Specific Distribution Service

Sites ordering Service C will receive one or more tapes containing pre-installed forms of Toolpack/1 (Release 2) software. All the major tools are included on each tape, though the experimental tools may not be. Source code for the supplied tools will also be provided. A copy of the Toolpack/1 Introductory Guide (Edition 2) will be provided with each tape. The range of tapes available under this service will increase as additional versions become available.

Service C.1 - DEC VAX/VMS. This tape includes compiled versions of the TIEVMS library, useful script files and executable forms (.EXE files) of all major tools. All relevant source files are also included. The software is compiled and linked under VMS version 4.3 and the tape is written in COPY format.

Service C.2 - Unix 4.2Bsd. This tar format tape contains a set of Toolpack/1 (Release 2) tools that provide Fortran transformation and analysis capabilities. A set of Unix shell scripts assist user access to tools and a Unix make file facilitates system installation. The tape includes documentation on the shell scripts.

Service C.3 - Apollo Domain. This WBAK format standard TAPE contains the main Toolpack/1 tools and all the supplementary libraries. This implementation is based on a tailored version of TIEC that allows command line arguments to be passed to the tools. The root of the PFS can be set by an environment variable. Both source and executable code is provided, plus build scripts for all tools and supplementary libraries. A small set of script files is included to aid in tool invocation.

Service C.4 - Apollo Domain. As Service C.3 except that the medium is Apollo standard CARTRIDGE.

Appendix B
Porting Toolpack/1 to IBM MVS/TSO

Martyn D. Dowell; Paul A. Moinil and Aurelio A. Pollicini
Joint Research Centre
Ispra, Italy

*"Reddite ergo quæ sunt Cæsaris, Cæsari;
et quæ sunt Dei, Deo."*

— Matthew, XXII, 21

Abstract. A case study is reported, which deals with the implementation of a front-end to the TIECODE Library Release 2, for hosting Toolpack/1 in an IBM environment under the Systems MVS/TSO.

1.   The Study Background

The Toolpack/1 software is intended to provide, among other facilities, *analysis and transformation of Fortran 77 application programs*. Therefore, in Toolpack/1 there are implemented such functions as scanning, parsing, indexing and their reverse unscanning and tree-flattening.

Toolpack/1 software was designed to be *portable over computer systems on which a standard conforming Fortran 77 processor is available*.

Is there not a subtle contradiction in the two statements above? To some extent there is, of course. How can you perform functions proper to a compiler and still be *portable?* In theory, this class of software can only be *portable* as a compiler could be. That is, it is *not portable* but it can be made *transportable* if:

(1) all system-specific functions are isolated and put into low level modules forming a non portable *kernel* , kept as small as possible;

(2) all system-independent functions encompassing that *kernel* are designed and implemented as a set of portable higher level modules;

A. A. Pollicini (ed.), Using Toolpack Software Tools, 311–319.

(3) a different *kernel* is implemented for each computer system which hosts the software package.

This is, indeed, what has been done and it is the merit of Toolpack designers to have kept the kernel reasonably small. The task common to Toolpack/1 installers is to provide a TIE conforming kernel on the host system. This task proved to be easy on systems which had served as bed for Toolpack design and, unfortunately, less easy on other systems.

Faced, in Summer 1986, with the implementation of Toolpack/1 Release 2 in a software environment consisting of the following IBM products:
- MVS/XA SP 2.1.3
- MVS/TSO E with ISPF/PDF
- VS FORTRAN Compiler and Library Rel. 1.4.1

we had to re-implement the kernel in an IBM specific form.

The consideration that most of the kernel, both in TIECODE and in our system-specific version, is coded in Fortran 77 is a tangible measure of how, and to what extent, the use of a standard programming language is a necessary condition for the development of portable software; but not a sufficient one.

2.   The Interface Model

Toolpack/1 software is distributed in a parametric form. This particular form allows for site customization. System-specific and site-specific values have to be supplied as customized macros and then the supplied preprocessor TIEMAC can generate a compilable version of the customized software.

The Tool Interface to the Environment (TIE) consists of a set of libraries. The version of TIE more suitable for adaptation to different hosts is the one entirely based on Fortran 77, known as TIECODE. TIECODE can be given four different shapes according to the choice of the libraries. Though the *embedded regime* is the only one providing full integration, we planned to install first the *stand-atop regime* under which the transfer of control between tools is not allowed.

Toolpack tools are TIE conforming and, therefore, they can run unchanged or with minor adaptations in the environment created by any correct implementation of TIECODE, provided that the same set of values is used for customizing both the libraries and the tools. Therefore, the effort required for the overall installation mainly concentrates on TIECODE, whilst tools are much easier to install.

In the particular frame of our task, a number of problems had to be solved. By their different nature they may be classified as:

(1) file structure and organization;
(2) dynamic access to files;

(3) extension of the standard set of Fortran Intrinsic functions;

(4) restrictions of the Fortran compiler in use.

The first point was essentially a problem of recovery of files from tape and of matching the host filing system. The logical tree-structured organization was dropped, except for the lowest level, flattened on the tape by packing sets of units, which was rendered by storing the individual units as members of as many partitioned files as the original packed sets. The second point was the crucial problem while the last two points implied additional site customization. The devised solutions are detailed in the following part of this section.

2.1.   The Mailbox Model

File handling is perhaps one of the less standardized topics in Standard Fortran. Referring explicitly to ANSI X3.9 – 1978 [1], we are reminded that:

" 12.2.1 File Existence. At any given time, there is a processor-determined set of files that are said to exist for an executable program. A file may be known to the processor, yet not exist for an executable program ... .

To create a file means to cause a file to exist that did not previously exist. To delete a file means to terminate the existence of the file.

... "

" 12.2.2 File Properties. At any given time, there is a processor-determined set of allowed access methods, ... for a file.

A file may have a name; ... . The set of allowable names is processor dependent and may be empty. "

Countless are the occurrences of the expressions "processor-determined" and "processor dependent", mainly referred to the specifications of "file-to-unit connection" and "input/output error conditions".

Most Fortran 77 processors intend files as dynamic objects and allow the user to create and delete them, physically, at run-time. Of course, this is the typical implementation provided by friendly systems which are intrinsically orientated towards conversational user interfaces. By the origin of its design, Toolpack implies such capabilities. In contrast with this quite general trend, in IBM VS FORTRAN, files are physical components of the periphery; only their contents may change at run-time, not their status. The "file name" is not the "name" of the "file", but rather a symbolic reference to a physical file which is established statically when the job is scheduled by the operating system. Moreover, when executing an "INQUIRE by file" statement, the specifier 'EXIST=' does not check the property of "file existence", not even the "processor-determined" concept of existence as intended and implemented in the VS FORTRAN processor. At least until the release we used, the specifier 'EXIST=' may recognize a file by name only after connection to a unit has been performed by the execution of an OPEN statement which contains the specifier 'FILE='. To carry our task through, we had to "normalize" the behaviour of VS FORTRAN to the requirements of Toolpack.

The necessity of handling files dynamically was approached as a procedural abstraction to extend the language with facilities available from the underlying systems. Since the conversational layer provided by TSO handles files dynamically, we devised to access the facilities we needed by creating a *virtual mailbox* into which the suitable TSO commands are dispatched by TIECODE units at run-time. The implementation of the mailbox model made possible, through a single primitive, to overcome most of the difficulties. File operations such as creation, allocation, rename, release and delete can be performed. (Tool scheduling at run-time becomes equally feasible and so the main problem posed by the embedded regime may have a self-contained solution.) File existence is checked against the contents of a supplementary *file attribute block* into which the actual status of all named host files is recorded and updated throughout a run.

2.2. The Front-end Model

Of the two approaches consisting of either embedding system-specific code wherever needed or providing an additional interface with minimal modifications to the original code, just for invoking suitable units from the interface, the latter seemed more consistent with the modularity and the hierarchical structure of Toolpack/1. Therefore, all modifications to the original TIECODE may be grouped in two classes: customizations and extensions. Only customization is performed directly within TIECODE units, while all extensions required by the implementation of the mailbox model are provided by a front-end library.

Among the important IBM-specific customizations are:

(1) Bitwise shift and logical operations are available intrinsics of the library which is part of the IBM Fortran processor. The functions ZIAND, ZINOT, ZIOR, ZLLS and ZLRS are adapted so that the IBM specific intrinsics are referenced.

(2) Time and date primitives (subroutine ZTIME) are provided by calling an assembler control section and then decoding the returned strings. This modification is site-specific and cannot be transported. However, a more general solution may be based on intrinsic subroutines available in the enhanced IBM processor VS FORTRAN Version 2.

(3) An IBM restriction to 25 parallel clauses in an IF-block, in some cases required that a single IF-block with many ELSE IF statements be split into two smaller IF-blocks.

Some technical details of the implementation of the front-end library are given in the next section.

3.   The Interface Software

The version of TIECODE implemented under the IBM systems MVS/TSO consists of a front-end library of eight program units and of a set of 200 TIECODE program units, 20 of which have been modified. On the top of the assembler unit TSOBOX – the essential component of the front-end – are built five functional primitives which simulate TIE conforming file access, plus two units for giving the whole simulation mechanism an automatic behaviour.

3.1.   The Engine TSOBOX

During a TSO session the Terminal Monitor Program (TMP) iterates over the steps of obtaining a user command, checking it for correctness and passing control to the command processor for that command. In the normal case, which is a conversational session running in foreground, the user command is input from the terminal keyboard. Less usual is the execution of the TMP in background (batch mode), in which case the user command is input from a host file. In both situations, however, user commands are input directly to the TMP. On the contrary the mailbox model implies that the user command to be processed by TSO is a character string being defined in storage by the Fortran program currently running under TSO control. TSOBOX obtains the user command as its first parameter and then simulates the remaining steps of command processing, by taking advantage of the surrounding TSO environment. Functionally, the main steps of TSOBOX are:

(1) check the parameters received by the calling Fortran subprogram;

(2) recognize the TSO environment and checks parameters for consistency;

(3) access the TSO service routine for scanning the user command;

(4) upon correct completion of the scanning, transfer the control to the appropriate command processor;

(5) check the completion codes of the command processor and return parameters to the calling subprogram accordingly.

How steps (2), (3) and (4) are performed is briefly described here. By invoking the EXTRACT macro instruction, TSOBOX obtains the Protected Step Control Block (PSCB) required later and may additionally recognize whether it is running in foreground or in background in order to reject commands not authorized in either of the modes. Using the address of the current Task Control Block (TCB), TSOBOX returns to the Originating TCB until the address of the work area of the TMP is reached. There the parameters of the current TSO session are available, and thus the consistency of the TSO environment can be verified. The environment being proved correct, TSOBOX simulates the behaviour of the TMP, by building a Command Scan Parameter List (CSPL) which contains the addresses of User Profile Table (UPT), Environment Control Table (ECT), Event Control Block (ECB), Command Scan Output Area (CSOA) and Command Buffer (CBUF). The TSO service routine IKJSCAN can then be invoked for scanning the user command contained in the buffer CBUF. The information returned in the output area CSOA is checked. If the command is syntactically correct, a Command Processor Parameter List (CPPL) is built and then control is passed to the appropriate Command Processor which performs the operations requested by the user command and returns the completion codes to TSOBOX.

It has to be added that, for a more general use than required by TIECODE, TSOBOX is also able to distinguish between TSO commands and command procedures. In the latter case the Command Processor EXEC is invoked for the processing of the command language CLIST.

3.2.   The Surrounding Structure

The way of conveying the TSO services provided by TSOBOX to the requirements of TIECODE implementation was managed by transmitting requests by file attributes. A file attribute block – a named Common block – is shared among front-end units. It provides room for 64 host files and for their connection to input/output units in the range 1:99. For each file four attributes are recorded:

(1) **Open-Close Status** which may be Scratch, Unknown, New, Old when opening a file or Keep, Delete when closing a file;

(2) **Access Mode** which may be Direct or Sequential;

(3) **Recording Format** which may be Formatted, Print-formatted or Unformatted;

(4) **Availability Status** which may be New, Old or List-file.

The entries of the file attribute block are accessed randomly by the hashing function QIFYRN. The mechanism for recording file attributes is set by the front-end subroutine QIFYON, called by the TIECODE unit ZINIT. The TIECODE library can then handle host files dynamically through invocation of three front-end units:

(1) the logical function QIFYCN referenced before OPEN statements (it occurs twice in the whole library);

(2) the logical function QIFYDC referenced after CLOSE statements (it also occurs twice);

(3) the subroutine QIFYIQ called in replacement of the INQUIRE statement (it occurs only once).

These front-end units encode a dynamic request in a *file attribute list* from which the appropriate TSO command is automatically generated by the character function

QJRCDR. The file attribute list has a length of 22 characters and presents different layouts according to the TSO command being generated. The first character identifies the command. The contents of the attribute list for the three essential commands is discussed briefly.

(1) The **Allocate** command enables access to a file. The second character specifies the file status which determines the action to be performed, i. e. either to access an existing file or to create a new one. Only in the latter case more file attributes are required in the subsequent fields:

   - Access mode (3rd char.);
   - Record format (4th char.);
   - Size of a physical record or block of records (char. range 5:8);
   - Size of logical record (char. range 9:12);
   - Space allocation unit (13th char.);
   - Quantity of space (char. range 14:16);
   - Name of disk volume – required whether site-specific restrictions allow user permanent files only on a pre-determined range of disks (char. range 17:22).

(2) The **Free** command disables access to a file. The second character specifies whether the file is identified by symbolic reference ("filename" in the IBM "sense") or by its physical host name. The third character may be used to request hard copy of a List-file.

(3) The **Delete** command destroys the file and immediately releases its physical space. No other fields are required in the attribute list.

The subroutine QJRCPL, which juxtaposes substrings ignoring trailing blanks, completes the set of units of the Front-end library.

4.   The Front-end Documentation

The first two files of the tape contributed to NAG Ltd contain the documentation which consists of a brief note describing the tape and of an installation guide. Some details concerning the tape description, which may be of general interest, are reported in this section.

The guide is specifically intended for installers of the implementation of TIECODE described in Section 3 above. Such a guide contains the directions to load the Front-end library, the pre-customized TIECODE library, the supplementary libraries and a utility library. There are detailed specifications of further customization that the TIECODE units ZINIT and ZTIME, as well as the utility MKPFS may require. Additionally, as an example of tool installation, it is shown how to load the lexer ISTLX from the "ISTLX.FILE" of the Toolpack/1 base tape and within the library environment just created.

4.1.   Customization and Target Characterization

The tape contains a version of the Toolpack/1 TIECODE software which is expected
to run on:

```
Machine ranges:      IBM  and  IBM compatible  Mainframes
Operating Systems:   MVS/370    or    MVS/XA
                     with  TSO  or  TSO/E
Fortran Processor:   IBM VS FORTRAN  (Rel. 1.3.0 or higher)
                     using the Optimization level 2.
```

The version is pre-customized as specified below:

```
implementation regime:    stand-atop
machine characteristics:  8 bits per character
                          4 characters per integer
                          0:255 bounds of character set
file store specification: IBM 3380 type disks
                          fixed blocked files
                          80 character records (text & data files)
                          9040 character blocks   "   "   "      "
                          7424 character blocks (PFS device files)
```

4.2.   Disclaimer

This software was contributed to, and is distributed by, NAG Ltd to provide assistance
to sites wishing to install Toolpack/1 Release 2. Neither the contributor, nor NAG Ltd
accept responsibility for the content, performance or suitability of this software and
neither will then be liable in any way for its use or the consequences of such use.

4.3.   Tape Format

```
Density:        1600 bpi
Tape label:     no header label
Data Code:      EBCDIC
Files 1 to 4:   record format is fixed blocked
                record length is 80 characters
                block length is 9040 characters
                (Sequential files copied by IEBGENER)
Files 5 to 12:  record format is variable blocked
                maximum block length is 9060 characters
                (PDS unloaded by IEBCOPY)
```

4.4.   Tape Contents

Twelve files are recorded on the tape. Mnemonic name and identification of the material contained in each file is listed below:

(1)	README	Tape description note
(2)	GUIDE	Installation documentation
(3)	TSOBOX.ASM	One assembler program unit
(4)	TIECODSQ.FORT	TIECODE (sequential stream of PU.s)
(5)	FRONTEND.FORT	Front-end for MVS/TSO
(6)	TIECODE.FORT	TIECODE (individual Program Units)
(7)	UTILIB.FORT	SPLIT, TIEMAC and Support Utilities
(8)	SUPPLIB.FORT	ACCESS, STRING and TABLES Libraries
(9)	TIE.DEFS	TIE Definitions
(10)	ACCESS.DEFS	ACCESS Definitions
(11)	ALIASES.DEFS	Additional File Definitions
(12)	PROC.CLIST	TSO Command Procedures

Files (4) and (6) are equivalent. Both contain 200 identical program units in the alternative file organizations: sequential and partitioned.

Bibliographic References

[1]   American National Standards Institute; "ANSI X3.9–1978. American National Standard
      Programming Language FORTRAN", *ANSI, New York*, (1978);
      also referenced as:
      "ISO 1539–1980(E). Programming Languages — FORTRAN" *International Organisation for
      Standardisation*, (1980).

[2]   American National Standards Institute; "S8(X3.9–198x). American National Standard
      for Information Systems Programming Language Fortran", *ANSI, Subcommitte X3J3, Draft
      S8, Version 104*, (June, 1987).

[3]   BAKER, B. S.; "An Algorithm for Structuring Flowgraphs", *JACM* **24**, (1977), 98-120.

[4]   BOYLE, J. M.; "Concrete Syntax and Semantics of TAMPR Transformations", *Technical
      Report, Argonne National Laboratory, Argonne, Illinois*, (1981).

[5]   BROOM, M. A. and HOPKINS, T. R.; "The Implementation and Use of Toolpack/1 on
      a Graphics Workstation", *Software – Practice and Experience* **17**, (1987), 561-573.

[6]   CLEMM, G. M.; "FSCAN Report and User's Manual", *Technical Report CU-CS-166-79,
      University of Colorado, Boulder*, (1979).

[7]   CLEMM, G. M.; "ODIN – An Extensible Software Environment: Report and User's Manual",
      Technical Report CU-CS-262-84, University of Colorado, Boulder, (1984).

[8]   COHEN, M. C.; "ISTPL/ISTPO Users' Guide", *Toolpack/1 Release 2 Technical Reference
      Manual: NP1296, NAG Ltd., Oxford, U.K.*, (1986).

[9]   COHEN, M. C.; "ISTYP – Toolpack/1 Parser", *Toolpack/1 Release 2 Technical Reference
      Manual: NP1300, NAG Ltd., Oxford, U.K.*, (1986).

[10] COHEN, M. C.; "ISTYF – Tree Flattener Users' Guide", *Toolpack/1 Release 2 Technical
      Reference Manual: NP1301, NAG Ltd., Oxford, U.K.*, (1986).

[11] COHEN, M. C.; "ISTAN – Execution Analyser", *Toolpack/1 Release 2 Technical Reference
      Manual: NP1304, NAG Ltd., Oxford, U.K.*, (1986).

[12] COHEN, M. C.; "TIE WINDOW Supplementary Library Guide", *Toolpack/1 Release 2
      Technical Reference Manual: NP1306, NAG Ltd., Oxford, U.K.*, (1986).

[13] COHEN, M. C. and ILES, R. M. J.; "Toolpack/1 Target Fortran 77", *Toolpack/1
      Release 2 Technical Reference Manual: NP1313, NAG Ltd., Oxford, U.K.*, (1986).

[14] Commission of the European Communities; "CEC SIC DPS13. Information Processing
      Standardisation Specification. CEC Requirements for Fortran", *CEC, Bruxelles*, (1984).

321

[15] COWELL, W. R.; "The Toolpack Tools ISTUD, ISTCD, and ISTSB: Guide for Users and Installers", *Toolpack/1 Release 2 Technical Reference Manual: NP1319, NAG Ltd., Oxford, U.K.*, also published as: *MCS-TM-74, Argonne National Laboratory, Argonne, Illinois,* (1986).

[16] COWELL, W. R.; "ISTLS Users' Guide", *Toolpack/1 Release 2 Technical Reference Manual: NP1321, NAG Ltd., Oxford, U.K.*, (1986).

[17] COWELL, W. R.; HAGUE S. J. and ILES, ROBERT M. J.; "Toolpack/1 Release 2 Introductory Guide", *Jointly issued as Publication NP1277, NAG Ltd., Oxford, U.K.*, and *Technical Report ANL-86-43, Argonne National Laboratory, Argonne, Illinois*, (Oct. 1986).

[18] COWELL, W. R. and SOBRINO M.; "ISTRF – Text Formatter", *Toolpack/1 Release 2 Technical Reference Manual: NP1318, NAG Ltd., Oxford, U.K.*, (1986).

[19] COWELL, W. R. and THOMPSON, C. P.; "Transforming Fortran DO Loops to Improve Performance on Vector Architectures", *ACM Trans. on Math. Soft.* **12**, (1986), 324-353.

[20] Digital Equipment Corporation; "Guide to Text Processing on VAX/VMS", *Manual AA-Y502A-TE, DEC, Maynard, Massachusetts*, (1984).

[21] DALY, C.and DU CROZ, J. J.; "Performance of a Subroutine Library on Vector Processing Machines", *in Vector and Parallel Processors in Computational Science, Duff I. S. and Reid J. K. [Eds.] – North-Holland*, (1985).

[22] DONGARRA, J. J. and EISENSTAT, S. C.; "Squeezing the Most Out of an Algorithm in CRAY FORTRAN", *ACM Trans. on Math. Soft.* **10**, (1984), 221-230.

[23] DONGARRA, J. J. and HINDS, A. R.; "Unrolling Loops in FORTRAN", *Software – Practice and Experience* **9**, (1979), 219-226.

[24] FICHEUX VAPNÉ, F. [Ed.]; "Fortran 77. Guide pour l'Écriture de Programmes Portables", *Éditions Éyrolles, Paris*, (1985).

[25] HAGUE S. J. and ILES R. M. J.; *Private communication*, (1986).

[26] HANSON, D. R.; "The Portable Directory System PDS", *Technical Report TR80-4, Dept. of Computer Science, University of Arizona, Tucson*, (March, 1980).

[27] HANSON, S. J. and ROSINSKI R. R.; "Programmer Perceptions of Productivity and Programming Tools", *CACM* **28**, (1985), 180-189.

[28] ILES, R. M. J.; "ISTED – Fortran Aware Editor Users' Guide", *Toolpack/1 Release 2 Technical Reference Manual: NP1274, NAG Ltd., Oxford, U.K.*, (1986).

[29] ILES, R. M. J.; "Toolpack/1 Command Executor", *Toolpack/1 Release 2 Technical Reference Manual: NP1276, NAG Ltd., Oxford, U.K.*, (1986).

[30] ILES, R. M. J.; "TIE Installers' Guide", *Toolpack/1 Release 2 Technical Reference Manual: NP1278, NAG Ltd., Oxford, U.K.*, (1986).

[31] ILES, R. M. J.; "Toolpack/1 Tool Installers' Guide", *Toolpack/1 Release 2 Technical Reference Manual: NP1279, NAG Ltd., Oxford, U.K.*, (1986).

[32] ILES, R. M. J.; "Tool Writers' Guide", *Toolpack/1 Release 2 Technical Reference Manual: NP1280*, NAG Ltd., Oxford, U.K., (1986).

[33] ILES, R. M. J.; "What is TIE?", *Toolpack/1 Release 2 Technical Reference Manual: NP1281*, NAG Ltd., Oxford, U.K., (1986).

[34] ILES, R. M. J.; "TIE STRING Supplementary Library Guide", *Toolpack/1 Release 2 Technical Reference Manual: NP1282*, NAG Ltd., Oxford, U.K., (1986).

[35] ILES, R. M. J.; "The TIE Routine Definitions", *Toolpack/1 Release 2 Technical Reference Manual: NP1285*, NAG Ltd., Oxford, U.K., (1986).

[36] ILES, R. M. J.; "ISTLX Users' Guide", *Toolpack/1 Release 2 Technical Reference Manual: NP1289*, NAG Ltd., Oxford, U.K., (1986).

[37] ILES, R. M.J.; "ISTAL Users' Guide", *Toolpack/1 Release 2 Technical Reference Manual: NP1303*, NAG Ltd., Oxford, U.K., (1986).

[38] ILES, R. M. J. and COHEN, M. C.; "TIE ACCESS Supplementary Library Guide", *Toolpack/1 Release 2 Technical Reference Manual: NP1283*, NAG Ltd., Oxford, U.K., (1986).

[39] ILES, R. M. J. and COHEN M. C.; "TIE TABLES Supplementary Library Guide", *Toolpack/1 Release 2 Technical Reference Manual: NP1286*, NAG Ltd., Oxford, U.K., (1986).

[40] JOHNSON, S. C.; "YACC: Yet Another Compiler-Compiler", *UNIX Programmer's Manual, 7th Edition Vol. 2B*, Bell Laboratories, Murray Hill, New Jersey, (1978).

[41] KERNIGHAN, B. W. and PLAUGER, P. J.; "Software Tools", *Addison-Wesley Publishing Company*, Reading, Massachusetts, (1976).

[42] METCALF, M.; "Effective Fortran 77", *Oxford University Press*, (1985).

[43] NAYLOR, C. M.; "Build Your Own Expert System", *Sigma Press*, Wilmslow, U.K., (1983).

[44] OSSANA, J. F.; "nroff/troff Users Manual", *UNIX Programmer's Manual, Vol. 2A*, Bell Laboratories, Murray Hill, New Jersey, (1979).

[45] OSTERWEIL, L. J. and CLEMM G. M.; "An Extensible Toolset and Environment for the Production of Mathematical Software", *Proceedings of the International Conference "Tools, Methods and Languages for Scientific and Engineering Computation"*, Paris, (17-19 May, 1983).

[46] SHAFTON, A. L.; COWELL, W. R. and OSTERWEIL, L. J.; "The Toolpack Fio Library Installers' Guide", *Technical Report ANL/MCS-TM-6*, Argonne National Laboratory, Argonne, Illinois, (Feb., 1983).

[47] TEITELBAUM, T. and REPS, T.; "The Cornell Program Synthesizer: A Syntax-Directed Programming Environment", *CACM 24*, (1981), 563-573.

[48] TRW; "Fortran 77 Analyser. Users Manual", *TRW, Defense and Space Systems Group*, Redondo Beach, California, (1984).

[49] WALLACE, D. R. and KUHN, R. D.; "Study of a Prototype Software Engineering Environment", *NBSIR 86-3408*, National Bureau of Standard, (June, 1986).

Index

E

F

G

H

I

N

O

P

T

U